PREFACE

It is evident that the market for manufactured goods and services is becoming both more global and more competitive. The experiences of leading international companies demonstrate that quality is a key attribute needed for competitiveness in the world market and is fundamental to market growth and profitability.

Electrochemical industries are no exception to this business trend. Papers in this symposium illustrate a few of the many applications of enhanced quality practices in the industry, including: control of electrolytic metal refining processes to reduce scrap and improve product quality; quality assurance in plating and application of coatings for corrosion protection in the automotive and food industries; development of ultraclean stainless steel gas delivery piping and electroplated parts for the semiconductor and electronic industries to improve functionality and reliability; and improvements in battery materials and batteries to satisfy the demand for higher capacity and reliability. To achieve these goals, the introduction and promotion of company-wide quality systems is indispensable.

It is our pleasure to introduce the twenty-one papers of this first Electrochemical Society symposium on Quality Management in Industrial Electrochemistry and to thank the authors for their contributions. The papers range from overviews of quality strategies and systems to detailed implementation of quality procedures in electrochemical research and manufacturing. They demonstrate that both the understanding and implementation of quality management have made great strides in the electrochemical industry. They also show that the pursuit of ever-higher levels of quality is hard work, requiring a deep commitment by the company, a sustained effort, and the execution of myriad details to assure that products and processes meet demanding quality specifications and customer expectations.

Finally, we observe that, while the quality management pathways pursued by the companies represented in this symposium may be different, they tend to converge on the same underlying ideas and strategies. The similarities outweigh the differences. Quality is a concept that transcends corporate, cultural, and national boundaries and has become a fundamental element in global commerce.

D. E. Hall and Y. Kondo

May 1993

TABLE OF CONTENTS

Process Control

Electroplating on Steel

Copper Electrolytic Processes

Other Processes

FACTS ABOUT THE ELECTROCHEMICAL SOCIETY, INC.

The Electrochemical Society, Inc., is an international, nonprofit, scientific, educational organization founded for the advancement of the theory and practice of electrochemistry, electrothermics, electronics, and allied subjects. The Society was founded in Philadelphia in 1902 and incorporated in 1930. There are currently over 6000 scientists and engineers from more than 60 countries who hold individual membership; the Society is also supported by more than 100 corporations through Patron and Sustaining Memberships.

The Technical activities of the Society are carried on by Divisions and Groups. Local Sections of the Society have been organized in a number of cities and regions.

Major international meetings of the Society are held in the Spring and Fall of each year. At these meetings, the Divisions and Groups hold general sessions and sponsor symposia on specialized subjects.

The Society has an active publications program which includes the following:

JOURNAL OF THE ELECTROCHEMICAL SOCIETY - The JOURNAL is a monthly publication containing technical papers covering basic research and technology of interest in the areas of concern to the Society. Papers submitted for publication are subjected to careful evaluation and review by authorities in the field before acceptance, and high standards are maintained for the technical content of the JOURNAL.

THE ELECTROCHEMICAL SOCIETY INTERFACE - INTERFACE is a quarterly publication containing news, reviews, advertisements, and articles on technical matters of interest to Society Members in a lively, casual format. Also featured in each issue are special pages dedicated to serving the interests of the Society and allowing better communication between Divisions, Groups, and Local Sections.

EXTENDED ABSTRACTS - Extended Abstracts of the technical papers presented at the Spring and Fall Meetings of the Society are published in serialized softbound volumes.

PROCEEDINGS VOLUMES - Papers presented in symposia at Society and Topical Meetings are published from time to time as serialized softbound Proceedings Volumes. These provide up-to-date views of specialized topics and frequently offer comprehensive treatment of rapidly developing areas.

MONOGRAPH VOLUMES - The Society has, for a number of years, sponsored the publication of hardbound Monograph Volumes, which provide authoritative accounts of specific topics in electrochemistry, solid state science, and related disciplines.

THE MALCOLM BALDRIGE NATIONAL QUALITY AWARD PROGRAM

Dale Hall* and Curt Reimann
National Institute of Standards and Technology
Gaithersburg, MD 20899 USA
(*Electrochemical Society Active Member)

The Malcolm Baldrige National Quality Award Program is a national effort to improve the quality awareness and performance of United States industry. The program gives the Malcolm Baldrige National Quality Award to the best U.S. companies as a recognition and reward for world-class excellence and provides information and guidance on quality principles, strategies, and methods to U.S. industry. This paper describes the quality philosophy and principles of the Baldrige Award program and reviews the program's experiences to date in evaluating, promoting, and rewarding quality performance.

INTRODUCTION

The Malcolm Baldrige National Quality Award Program is a national effort to improve the quality awareness and performance of United States industry. The program pursues this objective in two ways:

- The Malcolm Baldrige National Quality Award is given as a *recognition and reward for world-class excellence* to the best U.S. companies.

- The quality program provides *information and guidance on quality principles, strategies, and methods* to U.S. industry.

In this paper, we describe the quality philosophy and principles that form the foundation of the Malcolm Baldrige National Quality Award Program, the criteria that are used to evaluate and measure quality performance, and the program's scope and operation. We also review the program's experiences to date in evaluating, promoting, and rewarding quality performance.

A NATIONAL QUALITY PROGRAM - ORIGINS AND ORGANIZATION

In the early and mid-1980s, many industrial and government leaders realized that a renewed emphasis on quality was a key to improving the declining competitive position of U.S. industry. In response, the federal government created the Malcolm Baldrige National Quality Award under public law 100-107, the Malcolm Baldrige National Quality Improvement Act of 1987, to spur quality consciousness by rewarding companies that achieve a world-class standard of excellence. Although the Baldrige award is the most visible activity of the national quality program, the program's primary goal is the broad dissemination and application of quality values and precepts to improve the overall competitiveness of U.S. industry. Thus, the program is not only for the best quality performers in U.S. industry, but is designed to help companies at all levels of quality to perform better.

The award was named after Malcolm Baldrige, Secretary of Commerce from 1981 until 1987, in honor of his advocacy of quality management and his commitment to excellence in the federal government.

The national quality award program is a public-private partnership. The private sector helped to define, establish, and structure the program and funded an endowment for it through the Malcolm Baldrige National Quality Award Foundation. The private sector's active role was a key to the program's rapid acceptance by U.S. industry. The enabling legislation assigned management of the Baldrige program to the National Institute of Standards and Technology (NIST), the nation's standards and measurements laboratory, because of its world-renowned expertise in technical quality standards and its reputation for impartiality. The American Society for Quality Control (ASQC), an organization dedicated to the advancement of the theory and practice of quality control and allied arts and sciences, assists NIST under contract in administering the program.

A Board of Overseers appointed by the Secretary of Commerce provides advice and guidance to the program's management. Board members are distinguished leaders drawn from all sectors of the U.S. economy. The Baldrige program's Board of Examiners reviews all applications for the Baldrige Award. The more than 250 examiners on the board are quality experts from industry, government, academia, and trade and professional groups. Examiners are selected annually for one-year appointments in a competitive merit review. Once chosen, they are trained extensively on the award and its criteria. About 50 examiners are designated as senior examiners and 9 serve as award judges.

PHILOSOPHY OF THE BALDRIGE AWARD PROGRAM

The Baldrige award program is built on the premise that the drive to improve quality is a fundamental element of successful business practice. Furthermore, *the commitment to quality must be both pervasive and continual.* That is, companies must be dedicated to improving every aspect of their operations in the never-ending pursuit of excellence. As the principles discussed in this section make clear, the Baldrige program is not designed or intended to be a "quick fix" to the economic competitiveness problem in the United States. Rather, it is one element of a fundamental, systemic change in how American companies view and operate their businesses.

The Baldrige approach to quality recognizes that all organizations are different and must therefore develop quality systems that address their unique problems. Accordingly, the program is intentionally *non-prescriptive* and does not expound rigid methodologies for quality management. Instead, the award criteria specify the essential elements of a successful quality strategy. The Baldrige criteria thus constitute a framework on which companies build their own quality management systems. The criteria give companies wide latitude in designing and implementing their quality strategies -- the focus is on results rather than methods.

The Baldrige award program recognizes that honest and detailed *self-appraisal* is an essential first step toward achieving and maintaining high quality. The award criteria and guidelines were developed to select award winners but also, and more importantly, as self-analysis tools for all companies, whatever their positions on the quality curve. The criteria and the scoring system together comprise a diagnostic system that companies can use to assess their own quality performance and identify areas for improvement.

Self-appraisal and improvement depend on the Baldrige program's tenet that ***quality can be assessed and verified***. A quality theory based on facts and measurements allows practitioners to measure their progress against well-defined goals. Guesses and intuition are unsatisfactory means of devising plans and assessing improvement.

Finally, a substantial overall improvement in the quality performance of U.S. industry depends on ***sharing information*** on quality practices. The Baldrige program encourages companies to share quality expertise and serves as a national focal point for disseminating and exchanging information on quality management theory and practice.

Following its own maxims of self-appraisal and continuous improvement, the program continues to refine its core values and criteria. In 1992, for example, the program increased emphasis on the educational value of the award criteria to benefit the thousands of companies that use them in training, self-assessment, and design of quality systems. The scoring system was also adjusted to put more emphasis on results.

FUNDAMENTAL QUALITY VALUES AND CONCEPTS

The Malcolm Baldrige National Quality Award criteria are based on several fundamental values and concepts that are briefly summarized in this section.

Customer-driven Quality. The customer is the ultimate judge of quality. From the customer's perspective, quality embraces all product and service attributes that contribute value. Thus, companies need a strategic concept of quality and must strive for excellence in all customer transactions before, at, and after the point of sale. Customer-driven quality orients company thinking toward policies and practices that can increase market share and retain satisfied customers.

Leadership. The senior management of a business must assume the responsibility for quality leadership. Quality can become part of the corporate culture only with management's visible, personal commitment as leaders and role models. Corporate executives must create clear quality values, communicate them inside and outside the company, and build them into the way the company operates.

Continuous Improvement. The quest for higher quality must be unending. Companies must set ever higher goals for providing value to customers and improving their responsiveness and productivity. To achieve and maintain excellence, companies must engage in continual and regular cycles of planning, execution, and evaluation.

Employee Participation and Development. A company needs a fully committed, well-trained, and involved work force to achieve and maintain high quality standards. Consequently, the company has a responsibility to educate and train all of its employees and involve them in quality activities. Its reward and recognition systems should be structured to reinforce employee commitment to quality.

Fast Response. To be successful in today's rapidly moving economy, companies must shorten the cycles for developing and introducing new products and services. The company should simplify or revamp its internal procedures wherever possible to reduce response time. Actions that increase responsiveness often yield corresponding improvements in productivity and quality.

Design Quality and Prevention. Quality excellence derives from well-designed and well-executed systems and processes. The best companies build quality into their products, services, and processes and make corrective interventions as far upstream in their operations as possible. Attention to quality at the front end of operations can prevent process errors and eliminate defects in products and services, thereby improving both quality and productivity.

Long-range Outlook. Companies need clear, long-term quality improvement goals as well as strategic and operational plans to achieve sustained quality performance. They should develop long-term commitments to customers, employees, suppliers, and stockholders.

Management by Fact. To achieve their quality and performance goals, companies must manage their operations and make decisions based upon facts, reliable data, and analysis. Corporate management must develop, and judge its performance against, indicators that are significant, measurable characteristics of operations, products, and services.

Partnership Development. Companies should build partnerships that serve mutual and larger community interests. Such partnerships might include, for example, joint efforts with suppliers to elevate their quality performance, cooperation with customers in planning new products and services, agreements with unions to promote labor-management cooperation, and linkages with local educational institutions that supply the work force.

Corporate Responsibility and Citizenship. Companies should be good corporate citizens, demonstrating a high standard of business ethics and showing concern for public health, safety, and the environment. They should also be committed to sharing quality-related information in their business and geographic communities to promote a higher overall level of quality performance.

MEASURING QUALITY PERFORMANCE — THE AWARD CRITERIA

The philosophy and core values of the Baldrige program outlined above are embedded in the seven award criteria (1) described in this section. Each criterion defines an area in which companies must excel to achieve world-class quality. The primary emphasis in the criteria is on producing results. Thus, ***the Baldrige criteria are designed to link processes and outcomes,*** as indicated in Figure 1.

The weightings of the seven criteria, for evaluation as well as selecting award winners, are indicated by the point values assigned to each; the total point count is 1000.

1. *Leadership* (95 points)

   Senior executives must instill quality values into the organization. Their commitment to quality must be real and visible. Corporate leadership is responsible for defining clear and visible quality values, communicating the company's commitment to quality both inside and outside the company, and developing a management system to guide the company's quality efforts.

2. *Information and Analysis* (75 points)

   Quality improvement is a science, relying on data as other sciences do. In a quality system based on facts and data, the gathering and use of information is crucial. Therefore, companies must develop effective means of gathering and analyzing information on company performance using both internal data and comparisons/benchmarks against superior firms. As part of this effort, companies must develop quantitative indicators of quality performance and apply them to assess and improve performance.

3. *Strategic Quality Planning* (60 points)

Continuous quality improvement is a long-term competitive strategy. Quality must be an integral part of the total strategic business plan. Companies also need specific short- and long-range plans for achieving and maintaining quality leadership in their businesses. The company's goals must be clear.

4. *Human Resource Development and Management* (150 points)

The company must use the talents and abilities of its entire workforce with maximum effectiveness. Accordingly, the company must take steps to ensure that employees at all levels are trained, educated, and committed to the company's quality goals. Superior organizations first train and educate their workers and then empower them; they view upgrading employee competence as an investment in human capital rather than an expense.

5. *Management of Process Quality* (140 points)

The company must have systematic procedures for assuring the quality of its goods and services. Both process design and control are crucial elements of quality. Companies must emphasize prevention of problems rather than methods of handling problems after they occur. The company's quality control must also extend to products and services from its suppliers.

6. *Quality and Operational Results* (180 points)

Companies must demonstrate the effects of superior leadership and operations in terms of sustained and measurably improved quality. Improved results can include, for example, more efficient operations, higher quality products and services, and better relationships with customers and suppliers.

7. *Customer Focus and Satisfaction* (300 points)

As the final arbiters of quality, customers determine what a company must do and whether it has been done satisfactorily. Companies must show that they take pains to know their customers and their requirements and respond to them. They must demonstrate strong performance in customer service and a commitment to building long-term relationships with their customers.

Table I summarizes the criteria, the specific performance factors included in each, and the point values assigned.

The criteria, together with the scoring system, comprise a diagnostic system. The criteria define the essential quality requirements and the scoring system provides a means to assess performance against them. In practice, many companies that have not applied for the Baldrige Award use the criteria and scoring system for internal assessments as part of corporate quality improvement strategies.

The program recognizes and rewards quality excellence among U.S. firms each year with the Malcolm Baldrige National Quality Award. A maximum of six awards may be given each year: two in manufacturing, two in service industries, and two to small businesses (companies with no more than 500 full time employees). Since the Baldrige award was first given in 1988, nearly 400 applications have been submitted for the award and 17 awards have been given. Application and award statistics are shown in Table II.

Table II. Malcolm Baldrige National Quality Awards
Selected Data on Applicants by Category

		Manufacturing	Service	Small Business	Total
1988	Applications	45	9	12	66
	Site Visits	10	2	1	13
	Awards	2	0	1	3
1989	Applications	23	6	11	40
	Site Visits	8	2	0	10
	Awards	2	0	0	2
1990	Applications	45	18	34	97
	Site Visits	6	3	3	12
	Awards	2	1	1	4
1991	Application	38	21	47	106
	Site Visits	9	5	5	19
	Awards	2	0	1	3
1992	Applications	31	15	44	90
	Site Visits	7	5	5	17
	Awards	2	2	1	5

The awards are given for excellent quality performance according to the standards and criteria discussed in the preceding section. They are not given for specific products or processes, nor do they constitute endorsement of a company's product line or of specific products. When, as has happened, an award is given to a specific division or sector of a larger company, it pertains only to that part of the company that applied and was examined.

The Baldrige standards are "absolute," depending on the numerical ratings given in Table 1. Consequently, all six awards need not be given in any year. In fact, the program has made only 17 of 30 possible awards and has never made more than 5 awards in a given year, a reflection of its rigorous standards. In manufacturing, where the issue of U.S. competitiveness has focused, all of the 10 possible awards have been given. Service companies have earned 3 of 10 possible awards, and small businesses have won 4 of the 10 awards for which they were eligible.

The award process begins with submission of formal applications by companies that intend to compete for the award. (Only a small fraction of the companies that use the Baldrige award criteria for self assessment and training actually apply.) Applications then undergo a rigorous, four-stage review in which only those that pass one stage advance to the next.

Stage 1 includes reviews of each application by at least 5 members of the Board of Examiners. Based on these examinations, the Panel of Judges determines which applicants advance to the second stage. Stage 2 continues review of the applications by at least 5 examiners and 1 senior examiner. Significant differences in evaluations by individual examiners are resolved by discussion until a consensus is reached. Again, the Panel of Judges, using the results of these reviews, determines which applicants advance.

In stage 3, the examination increases in intensity as the reviewers turn from the written applications to site visits. As Table II shows, a relatively small fraction of the total applicants qualified for site visits in the program's first five years. At least 5 examiners and 1 senior examiner conduct thorough 3-5 day, on-site inspections of all stage 3 applicants to verify information contained in the written applications and to answer any questions that arise during the review process.

After site visits, the inspection teams prepare reports and submit them to the Panel of Judges. The judges conduct a final review in stage 4, focusing on the applicants' overall strengths and weaknesses. Judges consider applicants in each category one at a time, reducing the list until no more than two applicants remain in each category. The judges then vote on each applicant separately. They send the resulting recommendations to the director of NIST and, in turn, to the secretary of commerce, who selects the final winners.

The application and review process demand considerable commitment from applicants. However, they receive value in return in the form of detailed feedback reports on their strengths and areas for improvement as well as profiles of how they fared in the examination process. This information is useful, of course, to those companies that intend to compete for the Baldrige award again. However, its main benefit to all applicants is as an indicator of how well they are doing in implementing total quality and what they must do to achieve world class performance.

THE BENEFITS OF QUALITY

Quality is an important element of business success in a highly competitive marketplace. Quality performance can enhance a company's chances of higher sales, increased profits, and other desirable outcomes. External factors such as the general health of the economy, access to capital, strength of the competition, and governmental regulations may also have significant effects. However, quality excellence is a company's best means of positioning itself to exploit opportunities and meet competitive challenges.

The benefits of quality management appear in numerous ways. Extensive anecdotal evidence from the Baldrige Award winners and other leading U.S. companies and several studies by the U.S. General Accounting Office (GAO) (2) and others have shown that attention to quality can produce tangible benefits in all phases of company operation. The GAO study used as its industrial sample those companies that identified themselves as receiving site examinations in 1988 and 1989. In the remainder of this section, we illustrate the benefits of quality management using examples from the GAO study and elsewhere. Many more examples of specific quality benefits can be found in references 3-5.

Because quality is a systems concept, a strong quality program can streamline processes and operations throughout a company. In the GAO study, 59 of the 65 responses from 20 companies on operating indicators showed improvement as a result of quality management. Companies that instituted rigorous quality programs showed specific benefits such as reduced errors and defects (7 of 8 responses were positive, 1 showed no change), shortened product lead times (6 positive responses, 1 no change), and faster order processing (6 of 6 positive responses). Zytec Corporation, a 1991 Baldrige Award winner, reported a 26% reduction in manufacturing cycle time and a 50% reduction in the design cycle since 1988 (3). In the first ten years after beginning its "Pursuit of Excellence" campaign in 1981, Milliken & Company, a 1990 Baldrige award winner, increased productivity by 42% (4). A systematic approach to quality can also reduce waste and scrap in the manufacturing process: waste and scrap are not seen as inevitable, but as indicators of the potential for improvement.

Quality management can enhance the reliability and timeliness of delivery of products and services. All 12 companies providing data on reliability in the GAO study showed gains, averaging over 11 percent annually as determined by freedom from breakdown or error while in use by the customer. Xerox Business Products and Systems, a 1989 Baldrige award winner, decreased defects per 100 machines by 78% and unscheduled maintenance by 40% between 1984 and 1989 (5). Eight of 9 companies responding in the GAO survey reported improvements in the timeliness of their deliveries.

The use of facts and data in decision-making improves management efficiency. By gathering and analyzing data in assessing company performance using well-defined performance indicators, management is more able to make good decisions routinely. The expanded knowledge base also strengthens corporate planning and goal-setting.

Companies that commit to quality excellence reap rewards in enhanced employee pride, motivation, and morale. Workers respond positively to quality work environments through higher attendance, more attention to safety and health, and overall job satisfaction. In the GAO study, 18 companies reported improvements in 39 of 52 employee-related indicators. Increased employee pride and involvement result, in turn, in more and better employee suggestions to improve quality and productivity and to reduce costs. The companies reporting on this subject in the GAO study showed an average annual increase of almost 17 percent in the number of employee suggestions. At Westinghouse Commercial Nuclear Fuel Division, employee suggestions more than quadrupled between 1985 and 1988 (5).

Quality management, by emphasizing cooperative, mutually beneficial long-term relationships with suppliers rather than preoccupation with lowest cost, leads to more productive supplier/company relationships. Stable partnerships with suppliers enable the company to increase its exchange of information with them, so that suppliers can clearly understand and respond more quickly to company requirements. Cadillac Motor Car Company, a 1990 Baldrige award winner, increased the number of suppliers shipping on a just-in-time basis by 425 percent since 1986 (5).

Successful quality programs go beyond internal operations to focus on customer satisfaction. The GAO study of leading Baldrige Award competitors showed impressive gains in customer satisfaction as a result of quality management. Among 17 companies providing data in the GAO survey, 21 of 30 customer satisfaction performance indicators improved while only three declined. Improvements included overall customer satisfaction, reduced customer complaints, and customer retention. Xerox Business Products and Systems recently reported a 41% improvement in customer satisfaction with its products and services since 1985 (3).

Companies that have not made strong commitments to quality are often concerned about its cost. Excellent companies know, however, that quality pays dividends and that the cost of poor quality can be disastrous. The GAO study provides evidence that quality management can reduce the costs of developing and producing goods and services. The emphasis on designing for quality, that is, "getting it right the first time," prevents problems by building quality into products and services. As a result, the costs of inspection, rework, and warranties go down: all five companies in the GAO study that measured the cost of quality reported decreases averaging about 9% annually. Furthermore, all 9 companies responding showed cost savings from quality improvement suggestions ranging from $1.3 million to $116 million per year.

The ultimate goal of the pursuit of quality includes business success manifested as increased market share and profits. While no quality program can guarantee such an outcome, experience to date suggests that companies that achieve high levels of quality do, as a whole, improve their business positions. Using performance indicators of market share, sales per employee, return on assets, and return on sales, 15 companies reported 34 positive changes in those indicators and only 6 negative changes (2).

The benefits of improved quality are clear. However, given the rigors of applying for the Baldrige award, the detailed examination procedure, and the long odds against winning, why should a company participate in the Baldrige process? Baldrige award winners and other participating companies have said that *the primary value of the Baldrige program lies not in winning the award, but in the learning experience that results from competing* and going through the detailed internal and external analyses. Many companies that have competed without winning claim that they were well recompensed for their efforts by a better understanding of their strengths and weaknesses.

The Baldrige award also helps to prepare companies for success in international markets. With the increasing globalization of commerce, agreement on what constitutes quality benefits everyone. A step in that direction was taken in 1987 by the International Organization for Standardization (ISO) with its first ISO 9000 Standards, a series of quality management systems standards designed to ensure consistency of product quality and reliability. The ISO 9000 standards are part of a larger world-wide movement toward systematic and compatible quality standards. The more-comprehensive Baldrige program is not built on, patterned after, or affiliated with ISO 9000, but is compatible. Consequently, a company that performs well according to the Baldrige award criteria will be well positioned to achieve ISO 9000 certification.

IMPACTS OF THE BALDRIGE PROGRAM

Carrying the Quality Message to Industry

Publicity for the Baldrige program centers on the awards given each year to the nation's best companies. The award has been instrumental in focusing national attention on the program and quality management and in building the present high level of interest. However, the awards themselves, which recognize the quality excellence of a few firms, are dwarfed in importance by the program's main impact, the widespread dissemination and application of quality concepts.

As a focal point for the nation's quality movement, the Baldrige program is uniquely positioned to carry the quality message to industry and other sectors of American society. The program has developed extensive resources and networks for information transfer. Its main means of raising quality awareness continue to be the award criteria themselves, which are written in a way that makes them valuable both as training guides and self-appraisal tools for companies on the pathway to higher quality.

The usefulness of the award criteria has been widely recognized throughout industry, as shown by their rapid acceptance and enormous distribution. In 1988, the first year of the program, about 12,000 copies of the guidelines were distributed. Demand for the criteria has grown impressively; in 1991, the distribution was about 240,000 copies. Follow-up surveys by the Baldrige program office indicate that most of these copies were used as training and assessment guidelines by industrial firms, government agencies, schools, and health care facilities. Thousands of companies have studied and applied the Baldrige criteria to their operations. Many of them have instituted formal internal quality management programs based on the Baldrige award criteria. They include large, well known firms such as Texas Instruments, Rockwell International Corporation, Perkin-Elmer Corporation, Intel Corporation, Inland Steel Bar Company, FMC Corporation, and AT&T (3).

Companies that participate in the Baldrige program help to carry the message of quality to other companies. Many, including Motorola as a prominent example, have shared quality strategies with their suppliers and have moved to build strong, long-term relationships with them founded on quality. The Baldrige program expects award winners to share their strategies and teach other companies how to improve quality. In the five years since the award program began, Baldrige award winners have given more than 10,000 presentations to firms, industrial associations, and other groups. Other program participants have given a like number. Their message derives from the underlying conviction that quality matters and that companies can take definite steps to improve their own performance. The presentations by these quality leaders include specific information on their experiences and techniques that they have used to boost quality performance.

Corporate winners and others have also written extensively on the Baldrige program, quality strategies, and case studies of companies that have implemented and benefited from quality programs (6). In addition, the Baldrige program cosponsors the annual Quest for Excellence quality conference in Washington, DC, where award-winning companies share information on their quality strategies, programs, and results.

As a result of these efforts by the Baldrige program and its supporters in industry, communication on quality performance has increased substantially within and among companies. A common language of quality has been created to facilitate that communication. In addition, the program is generating a body of knowledge on quality that continues to grow and undergo refinement in the spirit of continuous improvement.

Diffusion of Baldrige Quality Concepts

The success of the Baldrige Award in spurring quality consciousness in U.S. industry has kindled interest in other sectors of the economy and society where quality is a key to enhanced performance.

The educational community has moved on quality on two fronts. Increasingly, colleges and universities have begun to teach quality management theory and practice, including lessons based on the Baldrige program's experience. Some business schools now require students to take course work in quality management. The U.S. educational establishment is also applying quality principles as it strives to regain the nation's position of educational excellence. The non-prescriptive, diagnostic, customer- and results-oriented nature of the Baldrige quality criteria are well suited to educational enterprises, where one of the principal challenges is linking process to results. Educational associations and individual school systems have come to the Baldrige quality program for help and have begun to experiment with quality programs of their own. Industry, recognizing that the key to its future competitiveness is a skilled workforce, has been heavily involved in quality initiatives in the educational community.

The health care industry, like education, has unique problems that can respond to a systematic and broad-based emphasis on quality. The industry is turning to the concept of quality as a means of improving services and controlling costs. Health care employers have begun to use the Baldrige criteria for training.

The Baldrige award has inspired a number of other quality awards. State and local governments have developed their own quality awards. More than 20 states, including New York, Connecticut, Wyoming, Minnesota, Colorado, and Maine (4, 5) have adopted or are developing formal quality programs, drawing heavily on the Baldrige principles and quality criteria.

Departments of the federal government have adopted their own awards to recognize quality. In June 1988, the Federal Quality Institute (FQI) was established to promote total quality management throughout the federal government. The Baldrige program, serving as a quality resource within the existing government framework, has shared information and cooperated with a number of federal agencies including the departments of Labor, Defense, Education, and Health and Human Services.

The Baldrige program has also made the United States a leader in the world-wide quality movement. Countries around the world have instituted quality awards, and many of them have consulted with the Baldrige program and studied its criteria and operations. For example, the European Foundation for Quality Management, established in 1988, studied the Baldrige award in creating the European Quality Award (5).

SUMMARY

The Malcolm Baldrige National Quality Award program is a national program to improve the quality awareness and performance of United States industry. Each year, the program recognizes and rewards U.S. companies that have achieved world-class excellence. In the five years since the program began, it has honored 17 companies with the Baldrige award.

The program also provides information and guidance on quality principles, strategies, and methods. The Baldrige award process is based on the premise that quality can be assessed and verified and stresses self-appraisal and continuous improvement as keys to quality. The non-prescriptive award criteria elaborate the elements of quality, which companies use as guidelines in developing their own quality systems, and against which they evaluate themselves.

Thousands of companies have adopted the Baldrige Award criteria for training and self-evaluation. The program has also influenced and assisted quality movements in other sectors of the economy and society.

REFERENCES

1. The Malcolm Baldrige National Quality Award Program Office, *1993 Award Criteria, Malcolm Baldrige National Quality Award,* National Institute of Standards and Technology, Gaithersburg, MD 20899.
2. United States General Accounting Office, *Management Practices: U.S. Companies Improve Performance Through Quality Efforts,* GAO/NSI-91-190, May 1991. This report presents original data and cites several other quality impact studies.
3. Bureau of Business Practice, *Award Winning Quality: Strategies from the Winners of the Malcolm Baldrige National Quality Award,* Prentice Hall, Waterford, CT, 1992.
4. C. L. Hart and C. E. Bogan, *The Baldrige: What It Is, How It's Won, How to Use It to Improve Quality in Your Company,* McGraw-Hill, Inc., New York, 1992.
5. Marion Mills Steeples, *The Corporate Guide to the Malcolm Baldrige National Quality Award,* ASQC Quality Press, Milwaukee, Wisconsin, 1992.
6. One good source of additional information is: D. A. Garvin, "How the Baldrige Really Works," *Harvard Business Review,* November-December 1991, pp. 80-93.

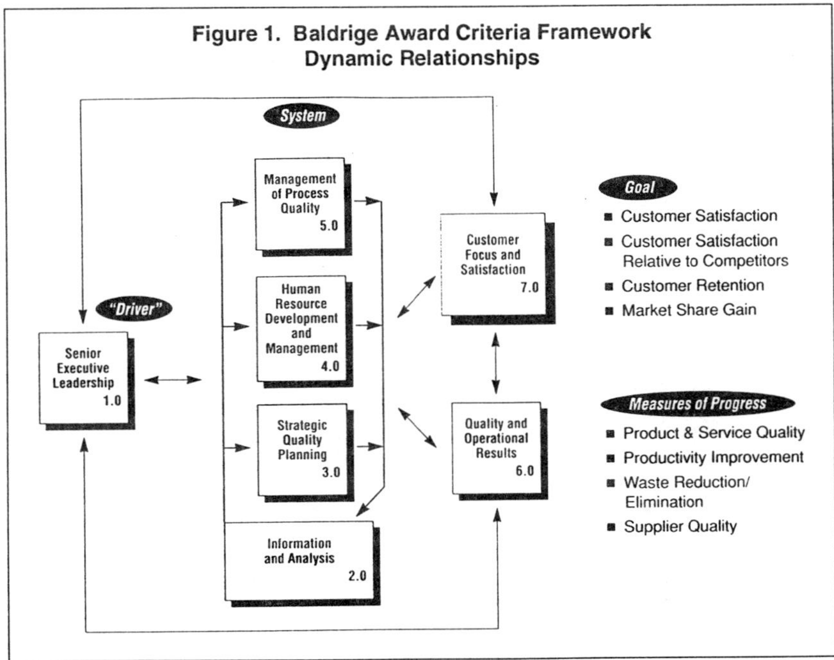

PREREQUISITES FOR WORLD-CLASS QUALITY

Yoshio Kondo
Professor Emeritus
Kyoyo University
29 Higashi-Takagicho, Shimogamo
Sakyo-ku, Kyoto 606 Japan

Quality is the key to competitiveness in the opening global market. The special features of quality of longer history and of common concern between manufacturer and customer make it more compatible with human nature than cost and productivity. When quality is improved in a creative way, cost is reduced, and productivity is increased. In addition to improving "must-be" quality, providing "attractive" quality is indispensable for exploiting the way to customer satisfaction. By doing this, the growth of market due to synergetic effect can be anticipated. By following the cycle of plan-do-check-act, not only the result of the work but also the process itself are improved in an upward spiral. Comparing the Japanese strategies with the lessons learned by the American Companies which won the Malcolm Baldrige National Quality Award, it was revealed that the way leading to the world-class quality is very similar.

INTRODUCTION

It is evident that the present decade of 1990s is much more demanding era for quality than what we experienced throughout the 1980s. The reason is that the present momentum toward the globally open and competitive marketplace has unstoppable force which no government nor regional business consortium can delay indefinitely anymore, even if they were inclined to do so. This will mean an enormous increase in the competitive pressures upon most companies.

The present experiences of the international leadership companies demonstrate that quality is the key to competitiveness in the market, that without this quality leadership, any products and service do not travel under any exclusive national passport, and that quality has become a fundamental way of managing any business anywhere for market growth and profitability. On the side of customers, it is an evident natural trend that their demands for quality continually increase along with the elevation of their living standard and educational level.

To cope with the competition in the global market, however, it is thought that not only the excellent quality but also the low cost of products and service and the high productivity of operations are the important and indispensable managerial elements. Why quality is most emphasized among them? To put this question into perspective, the following two points are emphasized. One is the special features of quality which distinguish it from cost and productivity, and the other is the harmonious relationship between quality and cost and between quality and productivity.

SPECIAL FEATURES OF QUALITY [1]

It can be said in the first place that the human desire for quality is not a new development, but the human history of quality is far longer than those of cost and productivity.

Table 1. Human History of
Quality, Cost and Productivity

Quality	1,000,000 years
Cost	10,000 years
Productivity	200 years

It is well-known that human being is an animal using tools, and it is thought that our ancestors had a keen interest in the quality of their tools. In the early centuries their major activities were hunting, stock raising and farming. All of these were aimed at providing food and clothing. The quality of first tools, such as arrowhead, plow and hoe, affected prehistoric man's catch and harvest. Our ancestors learned the importance of quality from their own experiences over a very long time since their appearance on earth more than one million years ago.

Compared with quality, the human connection with money is much younger. Self-supporting life had continued for a

long period during which people did not need to use money.
Along with the specialization of jobs and the development
of early cottage industry came the practice of bartering.
This further evolved into trade among adjacent villages and
among people who were far apart. Money was invented and
used as a convenient tool to catalyze transactions. The
concept of cost began to prevail much later - several thousands years ago, or, 10,000 years at the longest.

People started to discuss productivity during the industrial revolution about 200 years ago. The Taylor system originated only about 90 years ago.

As shown in Table 1, in comparing the human history of
quality, cost and productivity, quality is the longest.

It must be emphasized, on the other hand, that quality
is a major common concern between manufacturer and customer,
though they sometimes might have a different definition of
quality. In this regard, "customer satisfaction" or "fitness for use and environment" [2] from the viewpoint of customers is the most important concept in the company's activities of quality assurance.

Table 2. Main Concern of
Manufacturer and Customer

Manufacturer	Customer
Quality	Quality
Cost	Price
Productivity	After-Sales Service

In contrast to quality, customers do not have a keen
interest in cost; their primary concern is the price. While
the cost is determined by the conditions within the manufacturing company including the suppliers, the price is affected by the preference and demand of customers and by the
business fluctuation.

Goods are not sold simply because of higher productivity. The motive of the purchase might be that customers
can buy and repair it at any time and at any places. In

other words, they are persuaded by service performance after
the sale. The main concerns of the manufacturer and customer are summarized in Table 2. It is demonstrated that quality is the only common concern.

The desire for quality existed the longest, and quality is the common concern between manufacturer and customer.
These special features of quality make it more compatible
with human nature. It is due to this reason why the request
for quality improvement by the upper managers is more easily sympathized and accepted by the subordinates than the
call for cost reduction and productivity increase. Thus
the quality improvement is the most appropriate and acceptable way for enhancing corporate performance.

HARMONIOUS RELATIONSHIP BETWEEN QUALITY AND COST
AND BETWEEN QUALITY AND PRODUCTIVITY

There are opinions, on the other hand, that although
the importance of quality improvement is well understood,
cost increases and productivity decreases when quality is
improved. Then what we should consider is to look for the
optimum or balance between quality and cost and between
quality and productivity.

It is often asserted in the manufacturing process, for
example, that there be an optimum to quality of conformance,
or percent defective, with regard to manufacturing cost.
It is shown with solid line in Fig. 1 [3]. Some analysts
explain that increased conformance, or reduced percent defective, decreases the losses incurred by defects, but the
cost of quality improvement needed for greater conformance
rises sharply as quality approaches the perfect state.
Thus the optimum, or the minimal total cost, should always
fall short of perfection because the total cost soars as
the percent defective approaches zero.

However, the above optimum is doubtful; first, it ig-
nores the need of customers of which final goal is zero defect, and second, it does not consider the competition in
the market. If the competitor is successful in reducing
the manufacturing cost by reducing the percent defective,
for example, it is obvious that the name of this company
will disappear from the telephone directory sooner or later.

We should understand the difference in character of
the cost of quality improvement from those of basic manufacturing cost and losses incurred by defects. Both basic

manufacturing cost and losses incurred by defects are easily defined, and each of them is demonstrated with a single curve determined by the definition, respectively. On the other hand, the cost of quality improvement is usually indefinable: we know that there are always plural ways of improvement, and it is not shown with a single curve. When some creative ideas with which we can increase the conformance with less additional cost are introduced, the curve of quality improvement cost is shifted down as shown with broken lines in Fig. 1. Then the resultant total cost is lowered, and the optimum moves toward zero defect. If we could succeed to improve the product quality without quality improvement cost, the optimum is consistent with zero defect. Thus the optimum is movable and indefinable. What we must really do is not search for the indefinable optimum but to search for the ways and means with which we can improve the quality with minimum cost.

It may be said that the approach of this kind is a "breakthrough", which is quite different from the superficial optimization mentioned before. It is clear that the successful breakthrough is always accompanied by the creative idea and strong will of the people concerned.

Regarding the relationship between quality and productivity, Deming [4] told, "Productivity goes up as quality goes up. This fact is well-known, but only to a select few." This thought of harmonious relationship between quality and productivity is also based on the same idea of creativity, or breakthrough approach mentioned above.

Then it is summarized that when quality is improved in a creative way, cost is reduced and productivity is increased. It is seen that quality can be a cause of cost reduction and productivity increase, but low cost and/or high productivity do not always pave the way to quality improvement. It may be only logical, then, that we must start with quality whenever we attempt to improve a company's performance.

BACKWARD AND FORWARD QUALITY

It is important to note that Ishikawa [5] preferred to classify product quality as "backward" and "forward" qualities. Afterward, Kano [6] modified this classification to "must-be" quality and "attractive" quality, respectively, and demonstrated that they are mutually independent and that the classification should be two-dimensional.

Usual quality costs [7] only concern the must-be quality and are effective for reducing the failure costs caused by non-conformance, scrap, rework, custoner complaint, compensation, etc., although these quality costs are usually calculated within the company, and much of the costs due to poor quality paid by the customers is ignored. In the evolution of modern quality assurance methodologies, the early practice of 100 percent inspection (sometimes, several hundreds percent inspection) gave way to process improvements including designing to avoid manufacture of defective products. If this reduces defect levels to zero, the failure costs are remarkably reduced, and it solves the problem of short-term customer dissatisfaction but not necessarily the problem of customer satisfaction, or fitness for use and environment.

Providing attractive quality is indispensable for exploiting the way to customer satisfaction. By doing this, not only the improvement of market share but also the growth of market size due to synergetic effect are anticipated. As compared with must-be quality, attractive quality is easier to become latent and unnoticed by the customers themselves. In order to detect and grasp the hidden attractive quality, it is important for the manufacturer to collect and analyse the quality information from the market on the following items.

1. Customer demand, i. e. how the commodities are used and are convenient and inconvenient for the customers. The conditions of use.

2. Quality of similar commodities manufactured by the competitors.

3. Actual conditions of transportation and storage by distributors' channel and retailers.

4. Present and future market trend.

etc. The idea of hypothsis testing is effective in these surveys. The validity of the hypothesis that the handiness of camera is the attractive quality for the customers, for example, is tested by counting and comparing the number of people carrying small-size camera in downtown and in suburban areas [8]. The market survey of this kind should be carried out on the systematic basis.

In connection with the attractive quality, another problem of "surplus quality" which is mainly associated with higher quality of design should be discussed. A deli-

cate balance must be struck between achievement of higher
quality and the associated costs. Thus, conflicting pressures result from the desire to reduce costs and the desire
to elevate the technological level of the company and its
associated capability for new-product development. Since
the era of 1970s, Japanese television makers requested the
manufacturers of electronic components to reduce defect
levels below 10 parts per million. It took a great deal of
effort to meet this request, but the result has been beau-
tiful pictures on television sets, which are extremely reliable [2].

PDCA CYCLE

It is widely accepted in Japanese industries that the
control of process follows the so-called Deming's cycle,
which is composed of the four steps of plan, do, check and
act, as shown in Fig. 2(a) [2]. The objective (or standard) should be established and the process to attain the
objective should be given before doing the work. The results are then checked by comparing them with the target or
standard. Corrective actions of cause removal or standardization are taken when any significant difference is found,
and its causes are elucidated. By following this plan-do-
check-act (PDCA) cycle, it is expected that not only the
results obtained but also the process itself are improved
in an upward spiral. This may lead to improvement and
strengthening of the company's performance.

In some forms of manufacturing, the quality standard
and operation manual are established by the engineering
staff and managers, and the workers are only requested to
carry out their job of manufacturing in accordance with the
established manual. Thus the planning and execution are
separated. In such cases, if all the manufactured products
are found to be non-conformance, the supervisor seeks the
causes and may reproach the worker. The worker may then
reply, "I am not responsible for the defect. I honestly
followed the operation manual that you gave to me. You are
responsible for the results." It is clear that when workers
are responsible only for following the established manual,
their responsibility for quality becomes obscure. Such
vague responsibility is detrimental to high quality of conformance, which is achieved only if the workers are con-
scious of quality and have a keen sense of responsibility.

It is true that the workers are assigned to perform
the manufacturing job. However, this job performance is

20

also composed of a plan-do-check-act cycle, as shown in
Fig. 2(b). The extent to which PDCA cycle is followed in
this portion of the overall job is considered to reflect
the self-control ability of workers. Thanks to the ability
of self-control, we humans can enjoy our lives, including
sports and leisure. In order to cultivate the self-control
capacity of workers, education and training are the indispensable prerequisites.

This process of PDCA cycles is somewhat different from
the thought, "Do thing right the first time" which is pre-
vailing in the Western countries. We are afraid that no
one knows the right way of doing the work from the first.
The right way given in the operation manual is not always
correct and should further be improved. This is the reason why we emphasize and stick to the rotation of PDCA cycle
without cease.

STRATEGIES FOR WORLD LEADERSHIP QUALITY

Juran [9] gave his farewell lecture in Tokyo, October,
1989. In this lecture, he summarized the characteristic
Japanese strategies for world leadership quality as follows.

1. The upper managers take charge of quality.

2. The entire hierarchy is trained in how to manage for
   quality.

3. Quality improvement is undertaken at a revolutionary
   rate.

4. The QC circle concept enables the work force to participate in the quality revolution.

Supporting all of the above strategies has been the
adoption of the factual approach, he added, in which making
decisions is based on colletion and analysis of data, rather
than those based on opinions.

From the Japanese view, on the other hand, we have only
honestly studied and followed the thought and ways indicated
by Deming and Juran. We suppose that Japan is the only nation where the PDCA cycle is continually rotated very ear-
nestly. Perhaps the overriding national priority to over-
come the "quality crisis" since late 1940s has been far
stronger in Japan than in the Western countries. "If Japan
can, why can't we?" This is the title of a famous American
TV program in the middle of 1970s. On the contrary, the

Japanese mind has been occupied by a thought, "If America can, why can't we?"

In the same lecture in Tokyo, Juran also warned to Japanese that the Western countries are now in the process of adopting new strategies, some of which may well result in revolutionary improvements in their quality and that, in his view, the 1990s will become the decade in which the Japanese revolution in quality will encounter its first challenge.

Following this lecture, Juran [10] presented another paper "Strategies for World Class Quality" at the 34th EOQ Conference in Dublin, 1990. His major emphases in this address were as follows. The winning companies of the Malcolm Baldrige National Quality Award in the U. S. A. have made many stunning achievements in various fields within a few years. During making these achievements, the world-class quality companies learned a lot. The lessons they learned are summarized as follows.

1. Stretch goal can be met.

2. The Big Q concept must be adopted.

3. Clear ownership of multifunctional processes must be assigned.

4. An infrastructure for improvement must be created.

5. A lot of work is required.

6. Upper managers must personally lead the efforts.

7. The Taylor system must be replaced.

8. Quality goal must be incorporated into the business plan.

Comparing these lessons learned by the American companies with what are summarized as the Japanese strategies above, we can find a lot of similarities. Although the quality "climate" might be different among nations, the way leading to the world class quality is very similar. The fair competition and mutual cooperation are the important and indispensable conditions for attaining the world leadership quality.

SUMMARY

Quality has become a fundamental way of managing the companies for market growth and profitability, and the importance of quality is being emphasized in the present era of free market economy.

The human history of quality is far longer than cost and productivity, and quality is the only common concern between manufacturer and customer. These special features of quality make it more compatible with human nature.

When quality is improved in a creative way, cost is reduced and productivity is increased. Thus quality can be a cause of cost reduction and productivity increase. However, the opposite is not always true. It is only logical then to start with quality when we attempt to improve the company's performance.

In addition to improving the must-be quality, providing attractive quality is indispensable for exploiting the way to customer satisfaction. By doing this, the growth of market due to synergetic effect is also anticipated. The idea of hypothesis testing is effective in the market survey of attractive quality.

By following the cycle of plan-do-check-act, it is expected that not only the results but also the process itself are improved in an upward spiral. The importance of self-control of workers is emphasized to clarify their responsibility and bring up their humanity. The process of this PDCA cycle is different from the thought, "Do thing right the first time". We are afraid that no one knows the right way of carrying out the work from the first.

Comparing the Japanese strategies with the lessons learned by the American companies which won the Malcolm Baldrige National Quality Award, no significant differences are found. It is thought that the way leading to the world-class quality is very similar in various countries.

REFERENCES

[1] Y. Kondo, Quality Progress, 21, No. 12, 83 (1988)
[2] Y. Kondo, in "Juran's Quality Control Handbook 4th Edition", J. M. Juran and F. M. Gryna, Editors, p. 35F.1, McGraw-Hill, New YOrk (1988)
[3] Y. Kondo, Human Systems Management, 9, 7 (1990)
[4] W. E. Deming, "Erfaringer fra Kvalitetssyring I Japan", p. 87, Danish Society for Quality Control (1980)

[5] K. Ishikawa, "Introduction to Quality Control", p. 27, 3A Corporation, Tokyo (1990)
[6] N. Kano, N. Seraku, F. Takahashi and S. Tsuji, Quality, JSQC, 14, 147 (1984) (Japanese)
[7] A. V. Feigenbaum, "Total Quality Control Third Edition", p. 109, McGraw-Hill, New York (1983)
[8] T. Yoneyama, "Hinshitsu Kanri no Hanashi (Some Topics on QC)", p. 115, JUSE, Tokyo (1974) (Japanese)
[9] J. M. Juran, "The Evolution of Japanese Leadership in Quality", p. 25, JUSE, Tokyo (1990)
[10] J. M. Juran, Quality Progress, 24, No. 3, 81 (1991)

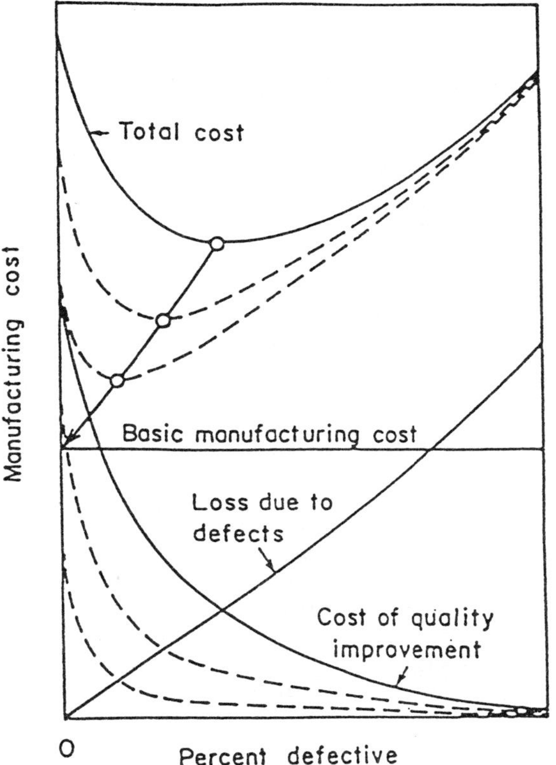

Figure 1. "Optimum"
of Manufacturing
Cost

Figure 2. PDCA Cycle

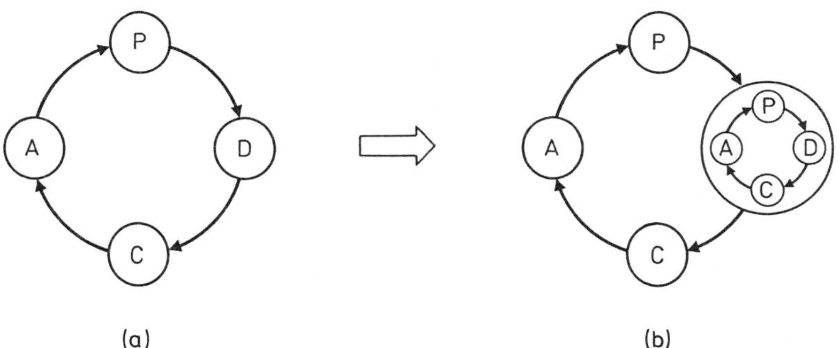

TQC Activities at NEC Kansai, Ltd.

Yoshihiko Muraki
Senior Vice President
Morio Katsuta
Senior Manager
Total Quality Control Promotion Division
NEC Kansai, Ltd.
9–1, SEIRAN 2–CHOME, OTSU, SHIGA, JAPAN

Abstract

NEC is a group of companies centered around a corporate philosophy geared towards the advancement of societies through its C&C concept.

NEC Kansai, as a member of NEC group, is responsible for the production of electron devices, including semiconductor components.

In 1984, a boom in the semiconductor market prompted NEC Kansai to increase its production. This, however, led to problems in product quality and equipment, adversely affecting productivity.

NEC Kansai, therefore, introduced the concept of Total Quality Control (TQC) in order to improve productivity and quality within the company.

Using the four "Pillars" described below as its foundation, NEC Kansai carries out its TQC activities based on its management philosophy.

there are

1)   Accomplishing work goals through policy management,

2)   Fostening quality assurance from the source,

3)   Enhancing productivity by improving facilities and raising facility design quality,

4)   Developing human resources through systematic education.

As a reward for all of these activities, we received the 1991 Deming Application Prize.

COMPANY OUTLINE AND HISTORY

NEC is a group of companies revolving around a common philosophy geared towards the advancement of societies through the C&C concept (Compute and Communicate). NEC Kansai has added to this its three-point management philosophy, mainly geared towards "Customer Values" based on QCDS, "Personal Dignity" in a management that makes best use of human talents, and "Technical Innovation" for revolutionizing business through technological advancements. Armed with this three-point philosophy, NEC Kansai continues to propel and excel in this business activities.

Last year, NEC Kansai registered sales of 132 Billion Yen, and currently employs 4,300 people. The pie-chart (Fig. 1) shows that semiconductors accounted for more than 55% of total sales, with the remaining percentage occupied by the other three business divisions. NEC Kansai's main office is located in Otsu City on the Southern Coast of Japan's largest lake, Lake Biwa. Located on the eastern side of Lake Biwa are four of NEC Kansai's branch plants.

NEC Kansai had its humble beginnings 50 years ago when NEC Corporation established a branch plant in Otsu City for the production of vacuum tubes. Forty years later, NEC Kansai became independent. We are now one of the production companies within NEC's Electron Device Group. From the moment of its conception, NEC Kansai strives to meet the needs of the times through technological improvements. It has always pursued this concept in revolutionizing its business activities, thereby contributing to its steady growth into what is now "NEC Kansai." See Fig. 2.

TQC INTRODUCTION AND PROGRESS

The concept of TQC was introduced at NEC Kansai in 1984. During that time, there was a boom in the electronic components market, and NEC Kansai's business performance was very promising. NEC Kansai, therefore, tried to increase its production, only to be crippled by problems with equipment and product quality. It was recognized that huge investments had to be injected into the semiconductor business that year, which therefore, called for faster recovery from the aforementioned problems.

NEC Kansai, therefore, introduced the concept of Total Quality Control (TQC), to build a stronger and a more fortified foundation for its business activities.

NEC Kansai focuses its attention on the four pillars that support TQC activities; namely, the accomplishment of business targets, the fostering of quality assurance through source control, the enhancement of productivity by improving existing facilities and raising the quality of facility design, and the implementation of an educational system for the development of human resources with excellent technical and problem-solving abilities.

During the first two years, NEC Kansai concentrated on educating its Managers for the improvement of their problem-solving abilities (1). In 1987, exactly three years after the introduction of TQC, NEC Kansai began its policy management. A year later, in 1988, NEC Kansai received the PM Plant Prize for the excellence of its TPM activities which were mainly intended in improving and strengthening its facility control.

In the latter half of 1989, NEC Kansai undertook a QC diagnosis for which findings showed that its number of weak points exceeded 507. NEC Kansai henceforth dealt with these points one by one.

For these series of activities, NEC Kansai once more proved its mettle by being awarded the 1991 Deming Application Prize.

MAJOR ACTIVITIES

Policy Management (Hoshin Kanri)

NEC Kansai's policy management is based on the basic three-point philosophy; "Carry out development in each division in line with the President's policy, follow the PDCA cycle with the participation of all employees, and establish a corporate structure that enables us to attain our business goals."

The following items were implemented for the promotion of policy management.

1)   Clarification of the main measures for the fiscal year based on mid-term plans

2)   Accurate policy execution through the use of forms

3)   Strengthening of implementation follow-up through the president's diagnosis.

These three items form the pillars that support NEC Kansai's policy management.

Fig. 3 depicts the policy management system of NEC Kansai. At the start of every fiscal year, the President reveals the main measures for the year which then make up the fiscal year policy. The division general managers then accept the president's policy and formulate their respective division policies. General managers and manager then accept the division policies for implementation and deployment. Managers formulate the concrete execution plans for execution of the policy.

After the policy is executed, results are checked monthly by division general managers and subordinates to hammer out measures necessary in accomplishing business targets. The president diagnoses the implementation of the policy twice a year, implements emergency measures that are found to be necessary, and carries over the diagnostic results to the next fiscal year.

NEC Kansai's management policy system was established not only for the purpose of achieving business targets, but also to act as the core of the company's TQC activities in order to ensure smooth promotion thereof.

<u>**Development of New Products**</u>

NEC Kansai develops new products based on the philosophy "Contribute to business growth by increasing sales through the development and provision of attractive new products that anticipate and satisfy customer needs." To implement this concept, NEC Kansai employs the QFD method, which proves to be the most effective.

The core of NEC Kansai's Development of New Products make up the following three items, namely, "Understanding and achieving required quality by the use of a quality table," "Improvement of design quality by strengthening the new product development system," and thirdly "Improvement of development ability by enhancing design support tools."

By using an example, how quality deployment was achieved and what kind of quality tables we prepared (2).

Fig. 4 consists of parts cut out from all the quality tables prepared from the development planning stage of a converter of a V-SAT outdoor unit (ODU), up to its design, trial production, preparations for production, and actual production. At the start of product planning, we tried to identify the quality needs of customers using a quality needs improvement table. The first quality table, which we called the "Quality Needs Deployment Table," was used to incorporate the quality needs of customers that we have identified, as characteristic values of products we manufacture. Using this table, we were able not only to check if the quality requirements of customers are fully satisfied by our product standards, but also to compare our products with those of other companies allowing us to identify prime selling points. Furthermore, we were able to identify bottlenecks in the manufacture of these products and make a "Technology Bottleneck Resolution Plan," and using this plan, we strived to solve these bottlenecks to assure that the entire development work would not be delayed. The next quality table, which we called the "Structural Deployment Table," was also made in order to implement the new product standards that we have formulated through quality needs deployment and to accomplish our design objectives. This quality table uses the quality characteristics in the quality needs deployment table as they are, and replaces the customer quality requirements with functional items, or in other words, the characteristics of entities in the development of products according to function. In this example, contents about electrical circuits account for almost two-thirds of the functional items for the accomplishment of design objectives and the mechanical structure accounts for the remaining one-third. Using this structural deployment table, we were able to see if the product design objectives and the design values of each part comprising the product match each other. In this example, cost development and major parts deployment are implemented, and at this stage, component drawings are prepared.

Next to structural deployment, reliability evaluation deployment and process quality deployment were executed using quality tables in the same manner. With reliability evaluation development, evaluation items are identified and conditions thereof are determined.

With process quality deployment, quality items are identified in each block of the manufacturing process and plotted with the quality characteristics of the new product on a matrix chart. This chart clearly shows the interrelation between the quality items and the quality characteristics, which help in determining manufacturing conditions and identifying process control items after manufacturing begins. Furthermore, in this example, the facilities and tools use in the manufacturing process are selected and control items necessary in maintaining the performance of these facilities and tools are identified. Here, information gathered from defects and problems occurring in past processes are also used. The control items that have been identified are indicated on the process control 4-unit set, like the QC process table, which proves to be of great use.

The results of the implementation of the QFD method in the development of the new product are the following. The number of design changes were decreased after the manufacture of the new product has started. In 1988 and the preceding years, there were 1 or more design changes made on an average. However, recently, since we are now able to develop new products that meet customer requirements right from the start, design changes has now basically become a thing of the past. Then the development lead time for new products were able to shorten. Although preparation of quality tables take time, development time has been shortened by approximately two months. Thanks to the QFD concept which has taken root in the company.

Intangible results were also obtained and this includes the improvement of communication with customers through the creation of a QFD concept, sharing of technical know-how among engineers in the company through paper forms, accumulation of technical know-how among engineers through educational materials that can be handed down from person to person, and improved joint work with related divisions through quality deployment. And, more than anything, the biggest result obtained is a deepened interest among leaders, supervisors, and managers in the engineering group to initiate improvements in work, on their own (3).

Quality Assurance

NEC Kansai's basic philosophy behind its quality assurance concept is "Provide reliable-quality products that anticipate customer quality needs, utilize the most advanced technologies, and achieve built-in quality at all quality assurance stages." NEC Kansai pursued its quality assurance activities based on this philosophy and the QFD method has proved to be a very powerful tool.

The fist pillar of major activities and progress is the "Quality assurance from the start of mass-production by improving evaluation activities." This forms the core of the design review method enhanced through quality deployment. In developing new products, design review is performed at each development stage before going on to the next. Even after the start of mass production, improvements to the quality and value engineering changes are common. Design review of the changes made is also performed, without fail.

This prevents any problems from occurring during initial mass production after the changes have been introduced. However, the phase from which design review is started is determined through the contents and extent of the changes, and FMEA (Failure Mode and Effects Analysis) is thoroughly conducted to obtain pre-production evaluation results.

The second pillar is "Built-in quality by improving process control." This forms manufacturing conditions and control items identified through quality deployment are integrated process control standards, and standard work is conducted based on these standards. This detail is described subsequently.

The third pillar is "Improvement of credibility by strengthening problem-recurrence prevention activities." Even before we introducted the TQC concept, we have established the use of forms like "Case Analysis Sheet," "Cause Analysis Sheet," and "Important Quality Problem Registration," as a way to prevent the recurrence of problems and claims. These were used to obligate the analysis of problems regarding quality, dig up the cause, and deal with the actual cause of trouble. More importantly, we made it sure that major problems in quality do not recur. As an end result, claims from customers decreased and customer satisfaction level improved considerably.

Now, let us go back to a more detailed explanation of the second pillar's process control. The objective is founded on the basic activity "Assurance of built-in quality in the production stage by complete process control." As I have mentioned in the explanation on the four major activities, manufacturing conditions and control items identified through quality deployment are integrated into the process control standards, and standard work is conducted based on these standards. Furthermore, we used control charts and improved the "4-unit Process Control Set" and the "4-unit Control Chart Set" based on the objective "Promote process stability." The four control charts are the "Instruction Sheet," the "Operating Condition Table," the "QC Process Table," and the "Equipment Inspection Standard." Quality requirements by the customer in the product developed through quality deployment were rewritten as control points in the manufacturing process. These charts are placed at the work place near the operators. By strictly observing these four charts, quality is built into products.

The "Instruction Sheet" clearly explains, through drawings, the important points in the work processes. The "Work Condition Table" is a list of the work conditions that correspond to each product, excluding the work procedures. The "QC Process Table" specifies the control and inspection items. The "Equipment Inspection Standard" designates the items to be inspected in tools, machines, and equipment. Information obtained from the process quality deployment in QFD are incorporated into the four process control charts and then employed in process control.

Equipment Control

The basic philosophy behind equipment control at our company is "To educate employees in the use of advanced equipment and increase productivity by improving equipment design quality and improving existing equipment." It goes without saying that, in the manufacturing industry, smooth operation of manufacturing equipment is of the utmost importance. Hence, we try to put our effort into equipment control through our TPM (Total Productive Maintenance) activities, and use quality charts when developing or purchasing equipment at our company.

We were confronted with two problems when we first introduced TQC. The first was poor design quality that eventually led to problems when equipment was operated. The second was an inadequate maintenance system that resulted in frequent equipment breakdowns thereby lowering productivity. Hence, we implemented various measures which I have summarized into two major points.

First, we introduced an equipment design review system employing equipment quality tables. We also conducted pre-production evaluation of equipment.

Second, we were able to raise equipment reliability by spreading and laying down the roots for self-maintenance, regular maintenance, and preventive maintenance through comprehensive equipment maintenance activities.

Let me give you an example of how production equipment achieved stable operation in a short amount of time as a result of conducting an equipment design review.

By breaking down the equipment into its characteristics and parts, we were able to identify technological bottlenecks. Next, we were able to forecast and improve reliability at the design stage by conducting equipment FMEA. We conducted design review along with the introduction of new equipment, and we prepared check sheets which enabled us to achieve our objectives.

Namely, we were able to improve purchase specifications through the use of equipment basic performance check sheets and prevent oversights in with-witness inspections, safety inspections, and acceptance inspection and tests. Moreover, we were able to argument know-how in equipment design by employing Equipment Design Standard Sheets and Equipment Information Feedback Sheets. As an end result, we were able to improve the quality of equipment designs, which ultimately led to the stable operation of production equipment in a short amount of time.

We were able to raise the success ratio of early-stage flow control and extend equipment MTBF.

OVERALL EFFECTS

So far, I have mainly discussed our TQC activities, especially our policy management and QFD. I would like to show you, now, the overall effects of our TQC activities, which I have compiled in Figures 5 to 8. From 1986 to 1991, we registered a 50% growth in sales which was one of our business targets. During the same period, we were able to reduce our loss-cost ratio to 1/6 through comprehensive quality assurance. In direct relation to increased productivity, sales per person during this period increased by more than 40% and total sales grew by 50%. We were therefore able to increase sales without increasing our work force, and this, I think can be seen as a direct effect of TQC.

Success in our personnel development and education led to a one and a half fold increase in the number of improvement proposals implemented in 1991 as compared with 65,000 in 1986. This is shown in Fig. 8.

Intangible effects were also obtained and these are:

1)   Business revolution continued revitalizing the company.

2)   Became a customer-oriented component supplier.

3)   Became more confident in the introduction of high-tech products and equipment.

FUTURE PLANS

Our vision is to transform into a company that can meet the demands of the 21st century, and this requires three things. First, we should offer and market products which prioritized customer-required quality.

Second, we should have a management system that makes the best use of people's talents, and to achieve this, the NEC group is currently conducting Super 21 Movement. Super 21 seeks to reevaluate systems and guidelines in order to improve business efficiency, revitalize the company's organization, and define the future vision for establishing a new corporate culture. We should train creative employees that can lead our business into the future, and to do this, we assembled 70 of our young employees, who probably will still be at the company in the coming century, and asked their vision of creating a new corporate culture.

Third, we should continuously improve our technical abilities. We have continued, through the past years, in pursuing the latest technologies, but I think it is also necessary to continue enhancing our own technology. It is also very important to widen our engineers' vision through academic activities and enhance our overall production technology. I consider production technology as anything that gives us the ability to easily produce designs and then automate those designs.

REFERENCES

(1)   K. Hosotani, "QC-teki Mono no Mikata·Kangaekata (in Japanese), JUSE Press Ltd., Tokyo (1984).

(2)   Y. Akao, (ed.), "Hinshitsu Tenkai Katsuyou no Jissai" (in Japanese), Japanese Standards Association, Tokyo (1987).

(3)   B. Kurahara, K. Uchimaru and S. Okamura, "Gijyutsu sya no TQC" (in Japanese), JUSE Press Ltd., Tokyo (1990).

Size
- Sales: ¥132.1 billion (fiscal 1991)
- Employees: 4300 (as of September 30, 1992)

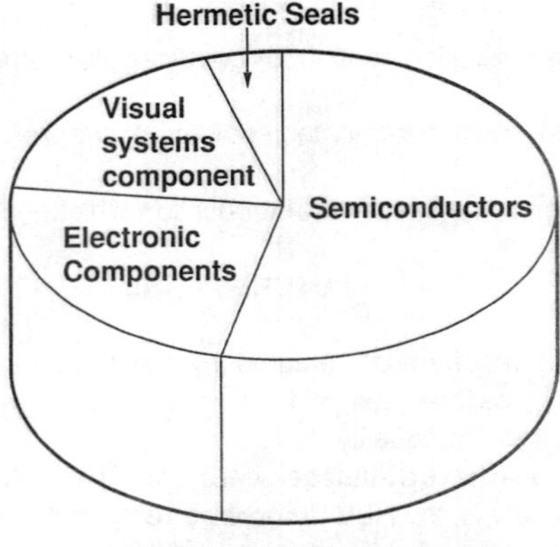

Fig. 1 Business overview

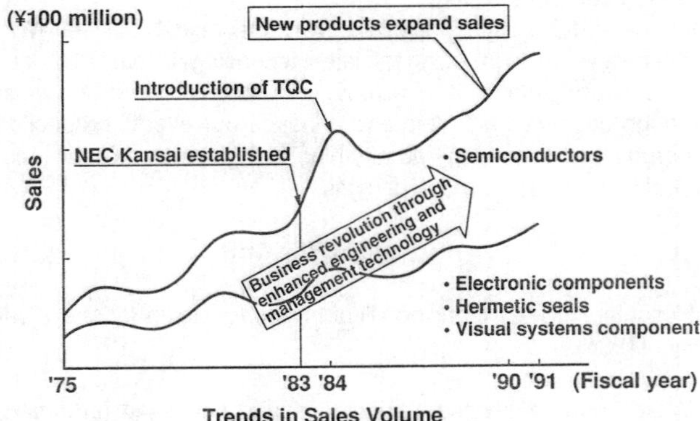

1943: Established as the NEC Otsu Plant

1953: Established Shin Nippondenki

1983: Established NEC Kansai

1984: Introduction of TQC

Fig. 2 History

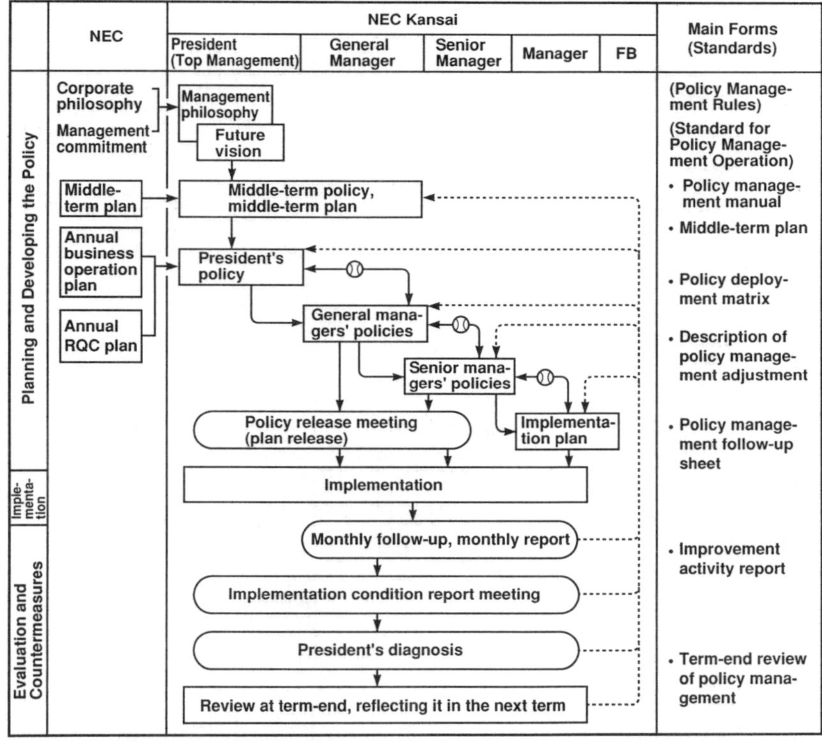

Fig. 3 Policy management system

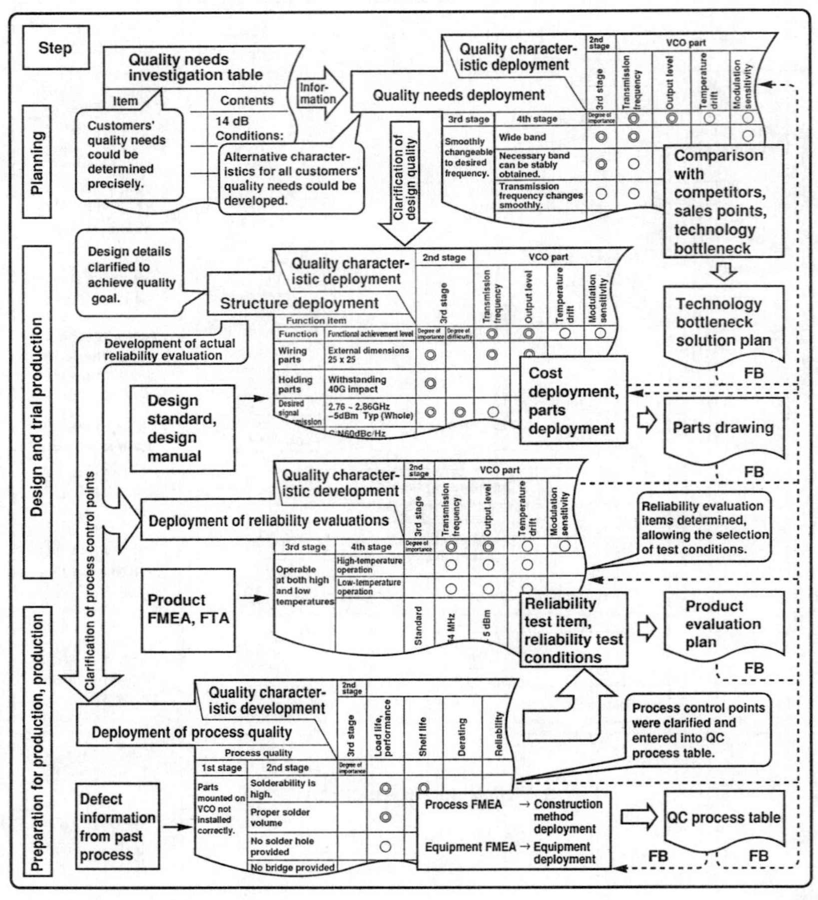

Fig. 4 Quality table for new product deployment system

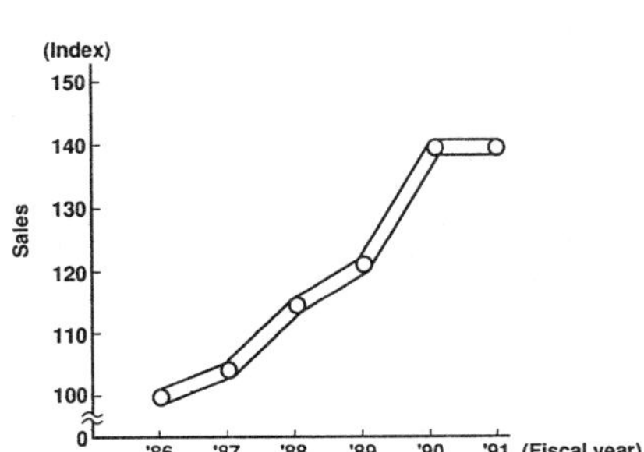

Fig. 5 Sales

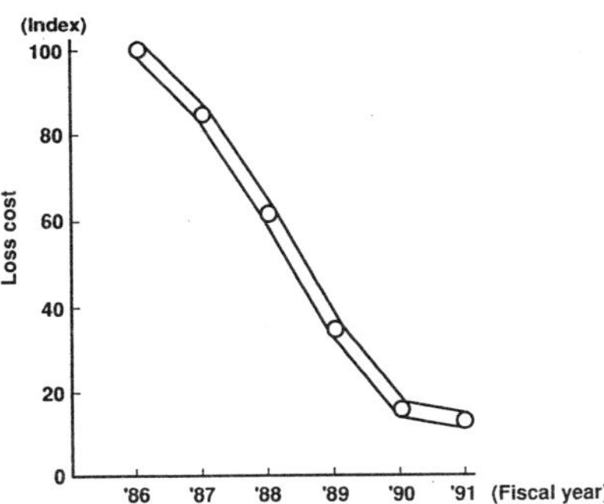

Fig. 6 Loss cost

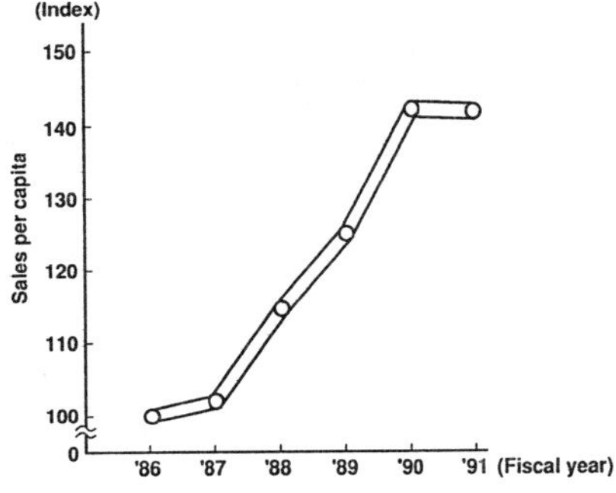

Fig. 7 Sales per capita

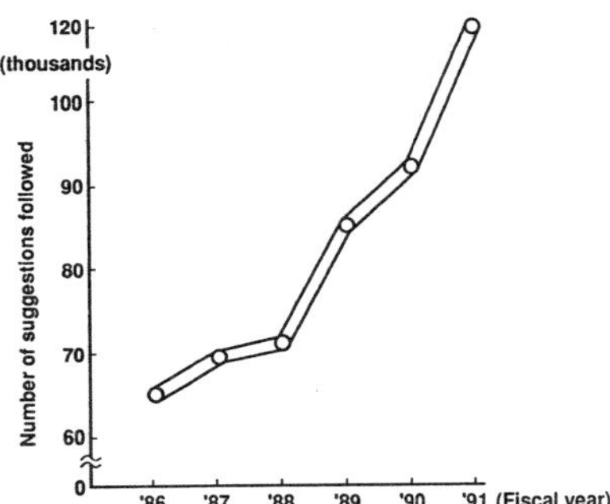

Fig. 8 Number of suggestions

QUALITY ASSURANCE ACTIVITIES IN NON-FERROUS METAL INDUSTRY

Tomoyuki Inami and Tetsuyoshi Kohga
Administration Dept.,Non-Ferrous Metal Div.
Sumitomo Metal Mining Co.,LTD.
Minato-ku, Tokyo, 105 Japan

ABSTRACT

The Quality Assurance Activities in Besshi-Niihama District Division of Sumitomo Metal Mining Co.,LTD. are introduced. Quality Assurance Section in Besshi-Niihama District Division was newly organized in August 1985 as the organization to manage and to drive the work of Quality Assurance collectively. Our Quality Assurance Section thinks three points important, which are Standardization, Quality Assurance Audit, and Quality Assurance Education in our company. The object of <u>Standardization</u> is to drive home to all members in Besshi-Division the management measures to raise control efficiency of general operation and to drive rationalization of the way of management. Internal <u>Quality Assurance Audit</u> is to audit every refinery and section by auditors from the Quality Assurance section to check Quality Assurance System. In <u>Quality Assurance Education,</u> we think repeated education important, and we are going to do so systematically.

INTRODUCTION

Many companies in Japan have actively introduced Total Quality Control(TQC). TQC is Quality Control with participation by everyone from the president to workers, and it is Quality Control with participation by members of every section, for example research,development, production engineering,manufacturing,purchasing,business, accounting,personnel,general affairs and so on. People from large enterprises to small-to-medium-sized enterprises, and from manufacturing industries to service industries,have made an effort eagerly aiming at TQC which is useful for management. Our company adopted completely the scienic management method as the method of rationalization of management in 1952, and decided to do

Company Standardization and Quality Control.

In 1982, we started the company-wide improvement activities which contained thought of TQC. We called them the FB(Fresh Besshi)Movement. By this, we have been trying to implement management by objectives and Quality Control education, prepare company standards, and activate small group activity together with making efforts to raise technical levels. Furthermore, in August 1985,the Quality Assurance Section was newly organized to manage and to drive the work of Quality Assurance collectively. The functions of the Quality Assurance Section are planning, making plans, drive, helping and guidance of the work of Quality Assurance, the audit for the company standardization, establishment and alterations and abolitions of company standards, the QA on development of new products, and other affairs on Quality Assurance.

The organization of Besshi-Niihama District Division is as follows. It has Toyo Smelter & Refinery, Niihama Copper Refinery, Nickel Refinery, Shisaka Plant, STF Niihama Plant, Electronics Niihama Plant,Isoura Plant, Construction & Engineering Center, General Affairs Sect., Safety & Environment Control Sect., Quality Assurance Sect., Physical Distribution Sect., Assay Sect., and others. This place is located in the center of our company which has developed as a totally non-ferrous metal maker. We also have a Copper Smelter and Refinery and Nickel Refinery and other facilities.

OBJECTIVES

In 1984, Besshi-Division of Sumitomo Metal Mining Co.,LTD. had many requests from our customers for new products and high purity products. So seven members who were chosen in various secitons in Besshi-Division investigated how the Quality Assurance System in Besshi-Division should exist. As the result, the points at issue were as follows.
1)Quality Assurance System is not arranged.
2)The Top Policy is not clear.
3)Quality Assurance Education is not done systematically.
4)Standards are not prepared.
5)The Correspondence of users' complaints is insufficient.
6)The Measure of the Estimation on Quality Assurance is
  not enough.
7)Internal Quality Assurance Audit is not done.
  Our objectives were to solve the above points.

RESULTS

After discussing the above points, the following proposals were made.
1)A Quality Assurance Section should be organized.
2)The functions of the Quality Assurance Section are as follows.
 a)Planning and Making Plans of Quality Assurance
 b)Drive, Helping and Guidance of Quality Assurance
 c)The Audit of Quality Assurance System
 d)Establishment, Alterations and Abolitions of Company
   Standards
 e)Management of Regulation of Complaint Handling
 f)Others
3)<u>A Quality Assurance Committee</u> whose chairman is the General Manager of Besshi Division should be formed. It should meet once per month to discuss and decide the fundamental rules and policies of Quality Assurance Activites.
4)<u>The Meeting of people in charge of Quality Assurance</u> should be formed as the subordinate organization of the Quality Assurance Committee. It should meet once per month to make plans of definite matters and study on the Quality Assurance Activities.
5)<u>A Quality Assurance Audit</u> should be done by an Auditor who is a member of Quality Assurance Section.

The above proposal was approved by the General Manager of Besshi Division, and the Quality Assurance Section was newly organized on 1st August 1985.
After organizing, the Quality Assurance Section has been driving three important items, that is <u>Standardization, Quality Assurance Audit and Quality Assurance Education</u> in order to perform exactly the above functions.

<u>1.Standardization</u>

This object is to drive home to all members in Besshi Division the management measures to raise control efficiency of general operation and to drive rationalization of the way of management.
The Keynotes on standardization are as follows.
1)The Standard must meet the actual condition 2)The Standard must be obeyed 3)The standard must be revised exactly to meet the actual condition 4)The Standard must be utilized.

40

Standards are as follows.
• Quality Standard
  Products Quality Standard(PQS)
  Materials Quality Standard(MQS)
  Business Materials Quality Standard(BQS)
  Intermediate Quality Standard(IQS)
· Quality Inspection Standard
  Products Inspection Standard(PIS)
  Materials Inspection Standard(MIS)
  Business Material Inspection Standard(BIS)
  Intermediate Inspection Standard(IMS)
• Operation Standard
  Engineering Standard(ES)
  Standard of Operation Procedures(SOP)
• Maintenance Standard
  Standards of Arrangement Members(SAM)
  Standards of Maintenance Procedures(SMP)
  Besshi Design Standard(BDS)
• Assay Standard
  Sampling Method Standard(SMS)
  Analytical Method Standard(AMS)
• Physical Distribution Standard(PDS)

<u>The Check of Standards</u>

Standards of every refinery and section are checked by members of the Quality Assurance Section once per two months.
Check points are as follows.
1)Are there differences between the standard and the actual work? What is the reason, if there are differences?
2)What is the point at issue in the actual results of control items?
3)What are other points at issue?
The Quality Assurance Section guides every refinery and section to revise the standard concerned more than once per year if there are differences between the standard and the actual work.

2.Quality Assurance Audit

In Quality Assurance Audit, there are Internal Quality Audit and External Quality Audit.

<u>Internal Quality Audit</u>

The Internal Quality Audit is to audit every refinery and section by auditors from the Quality Assurance Section <u>to check Quality Assurance System.</u>
The content of the Internal Quality Audit is to check once per three months how complaint handling and standards do as they are decided accurately in the refinery or section concerned.

Comparing operations actually done with the standard,the
Quality Assurance Section leads the refinery and section
to reform standards and to write down records accurately
if they didn't do as they decided.

<u>External Quality Audit</u>
Customers in Japan carry out the Quality Assurance Audit
in order to check the Quality and Quality Assurance
System of their suppliers.
When the result of Quality Assurance is bad, advice to
improve or halt operations is made. The instance of Check
List of External Quality Audit is as follows.

·1.The Preparation Condition of Standards :
 (1)Is the Preparation of Standards(Products,Operations,
    Inspection, Machine and so on)enough?
 (2)Are the Procedures of Establishment and Amendment of
    Standards appropriate? And is everybody available in
    the Standards?
 (3)Is the Quality Control System Chart and Work
    Responsibility Schedule clear?
·2.The Control Conditions of Facilities and Measurng
    Instrument :
 (4)Is the Maintenance of Facilities and Machines done?
    Is it done to keep records?
 (5)Do you have the Control Amendment System about the
    Measuring Instruments and Tools? Is it done to keep
    records?
 (6)Do you have special inspection facilities and
    apparatus for special use?
·3.The Control of Work
 (7)Is the Indication to New Work and Special Work
    appropriate?
 (8)Is the Contact in advance of the Improvement and the
    Amendment of working method thorough?
 (9)Do you grasp the Non-Conformity in the Process after
    carrying out the Intermediate Inspection?
 (10)Is the Indication of Marking of Conforming Product,
    Non-Conforming Product and Reworking Product clear?
 (11)Is the Example of Accidents which occured so far
    arranged and put to practical use by workers?
 (12)Is the Process Control for the Delivery Date done?
 (13)Is the method of storage,rust-prevention,
    moisture-prevention, and dust-prevention appropriate?
 (14)Is the method of Packing and Transportation enough?
·4.Quality Control
  ‹ QC System
 (15)Is there a Policy and Standard which indicates the
    Responsibility and Authority for Quality Control and
    Inspection?
 (16)Is the Quality Assurance System of Products and the
    Responsibity System for Quality clear?
 (17)Is there a Special System? Is the Responsible Person
    of Judgement decided?

(18) Is the Control Chart and the Statistical Method by
     means of Quality Control put to practical use?
(19) Is there a System which treats Complaints from
     customers? Is defect correction carried out speedily
     and precisely?
(20) Is the Sampling inspection plan appropriate? Is this
     done as it is decided?
(21) Is the Retention of Certificates of Analysis and
     Document of Inspection enough?
(22) Is there a system to audit internal conditions of
     Quality Control?
     Purchasing Control
(23) Is the Quality Control of Purchasing and Outside
     Product carried out ?
(24) Is the Method of Quality Identification (Acceptance
     Inspection) of Raw Material appropriate?
(25) Is the Division of storage of Raw Material clear?
(26) Are causes and phenomena of defective purchases put
     to practical use as the Statistics of Defects?
(27) Is the Purchasing Specification arranged and is the
     content completed?
(28) Is the System of the Corrective Action Procedure of
     Purchasing Product established and is the guide
     appropriate?
5. Others
(29) How about the Plan and the Operation of Training for
     employees?
(30) Is the Training of Workers(Working Attitude,Teaching
     Manners, Skill etc.)carried out?
(31) How about Safety Control of Work and the Condition
     of Environmental Protection?
(32) Is the Policy on the Safety of Products and the
     Consideration to Customers appropriate?

3.Quality Assurance Education

In Quality Assurance Education, we think repeated
education important, and we are driving it in two parts,
that is External Education and Internal Education. We
dispatch our members in various class to External
Education.
The Content of External Education is shown in Table 1.
In Internal Education, we are going to use a
systematic approach. The Content of Internal Education is
shown in Table 2.
The Content of Key Men Education
Quality Control is carried out using QC Seven Tools, that
is 1)Parato Diagram 2)Cause and Effect Diagram 3)Graph
4)Check   List   5)Scatter   Diagram   6)Control   Chart

7)Stratification
<u>The Techniques of Medium or High Grade</u> are as follows.
1)Control Chart 2)Statistical Test and Estimation
3)Correlation Analysis 4)Regression Analysis 5)Analysis
of Variance 6)Design of Experiment 7)Analysis of Multiple
Correlation

CONCLUSIONS

By driving the above Quality Assurance Activities,
we have gained many benefits, such as activation of the
workshop, increased Quality-consciousness, decrease of
customers' complaints, maintaining customers' trust, and
cost-reduction, as well as benefits which can not be
evaluated in monetary terms.

Table 1. The Content of External Education

Time	Theme	Sponsorship	Participator	Content
Apr. 1991 ~Sep. 1991	Quality Control Basic Course	Union of Japanese Scientists and Engineers(JUSE)	2 persons	6months(5days per month) Osaka Course
Apr. 1991	Seminar for Quality Control Basic Course	Japanese Standards Association(JSA)	3 persons	To master the Basic thought and Technique of Quality Control
Jun. 1991	Ditto	JSA	1 person	Ditto
Jun. 1991	Seminar for Total Quality Control Promotion	JSA	2 persons	Lecture and Group Discussion about the necessity and thought of TQC
Aug. 1991	Seminar for responsible person for industrial standardization and quality control promotion	JSA	2 persons	3days(short-term special training course)
Sep. 1991	Seminar for level up of Quality Control	JSA	1 person	Seminar for the level up of application ability in the actual workshop as the object of men of experience of QC
Oct. 1991	Seminar for New QC Seven Tools	JSA	5 persons	3days(short-term special training course)
Nov. 1991	The QC Annual Conference for Foreman	JUSE	11 persons	Participators join Special lecture Announcement, Disucussion and so on
Nov. 1991	The QC Annual Conference for Managers	JUSE	2 persons	Ditto
Nov. 1991	The Special Lecture Meeting of Quality Month	JSA	5 persons	The Lecture Meeting by Authorities of Quality Control
Nov. 1991	Seminar for Quality Control Basic Course	JSA	5persons	To master the Basic thought and Technique of Quality Control

Time	Theme	Sponsorship	Participator	Content
Feb. 1992	Seminar for Quality Control Basic Course	JSA	2 persons	To master the Basic thought and Technique of Quality Control
Mar. 1992	Seminar for Internal Standardization	JSA	1 person	Internal Standardization of JIS Factory, the theme of QA and the efficiency of production activities
Mar. 1991	146th Issuikai Quality Control Study Meeting	Issuikai Quality Control Study Meeting (The manager of this meeting: MEIDENSHA)	1 person	Study Meeting(Theme: Activation of Small Group Activities)and the tour through the factory
Jun. 1991	147th Issuikai Quality Control Study Meeting	Ditto	1 person	Study Meeting(Theme: The Control of Development Products and First stage variable) and the tour through the factory
Sep. 1991	148th Issuikai Quality Control Study Meeting	Ditto	3 persons	Announcement on QC Circle Activities and the tour through the factory
Nov. 1991	149th Issuikai Quality Control Study Meeting	Ditto	1 person	Study Meeting(Theme: The Quality Control of many grades and Small Quantity)and the tour through the factory

Table 2. The Content of Internal Education

Time	Theme	Sponsorship	Participator	Content
14th, 17th Feb. 1986	QA Education (Superintendent Course)	Quality Assurance Sectoin	35 persons	Presentation and discussion on Quality Assurance. Discussion Theme is "Pick up the points at issue on Quality Assurance and find the solution to the problem as the superintendent. "

26th Feb. 1986 ~ 30th Apr. 1986	QA Education (Foremen Course)	Quality Assurance Section	205 persons	Presentation and discussion on Quality Assurance. Discussion Theme is "Read a Quality Assurance Story written down by Quality Assurance Section and then pick up the points at issue on Quality Assurance and find the solution to the Problem as the foremen.
Apr. 1987 ~Sep. 1987	QA Education for Key Men (The first term)	Quality Assurance Section	15 persons	Participants must find the solution to the problems which they select from the points at issue on Quality Assurance existed in their workshops within six months.
Oct. 1987 ~Mar. 1988	QA Education for Key Men (the second term)	Quality Assurance Section	12 persons	Ditto
Apr. 1988 ~Sep. 1988	QA Education for Key Men (the third term)	Quality Assurance Section	10 persons	Ditto
Oct. 1988 ~Mar. 1989	QA Education for Key Men (the fourth term)	Quality Assurance Section	8 persons	Ditto
Apr. 1989 ~Sep. 1989	QA Education for Key Men (the fifth term)	Quality Assurance Section	10 persons	Ditto
Apr. 1991 ~Sep. 1991	QA Education for Key Men (the 9th term)	Quality Assurance Section	8 persons	Participants must find the solution to the problems which they select from the points at issue on Quality Assurance existed in their workshops within six months. And Training for Technique of Quality Control.

Oct. 1991 ~Mar. 1992	QA Education for Key Men (the 10th term)	Quality Assurance Section	11 persons	Participants must find the solution to the problems which they select from the points at issue on Quality Assurance existed in their workshops within six months. And Training for Technique of Quality Control.
Apr. 1992 ~Sep. 1992	QA Education for Key Men (the 11th term)	Quality Assurance Section	9 persons	Ditto
Oct. 1992 ~Mar. 1993	QA Education for Key Men (the 12th term)	Quality Assurance Section	12 persons	Ditto
Apr. 1993 ~	QA Education for Key Men (the 13th term)	Quality Assurance Section	14 persons	Ditto

ELECTROLYTIC MANGANESE DIOXIDE QUALITY MANAGEMENT
A CASE STUDY

Richard F. Wohletz and **Everette M. Spore**
Kerr-McGee Chemical Corporation, P. O. Box 55
Henderson, Nevada 89009

Morris P. Grotheer and **Samuel F. Burkhardt**
Kerr-McGee Technical Center, P. O. Box 25861
Oklahoma City, Oklahoma 73125

This case study of electrolytic manganese dioxide (EMD) quality management covers a period of time when competition among alkaline battery producers raised quality standards for EMD. Improved production technology was developed to meet the challenge of higher quality and greater uniformity while maintaining low production cost. The concepts of Total Quality Management focused on continuous improvement for the benefit of customers.

This case study is a learning experience in both production technology and quality management. Quality Action Teams were formed to address specific problems. The quality improvement program produced several patents on titanium anode designs and cell feed preparation. New methods for measuring the most important EMD quality characteristics were developed.

INTRODUCTION

Electrolytic   manganese dioxide, EMD, MnO_2, is widely used as the cathode material in "dry cells". The first EMD was used as a supplement to natural MnO_2 ore in Leclanche' cells, later in alkaline cells and most recently in various lithium batteries. This paper describes the historical striving for EMD quality at Kerr-McGee Chemical with emphasis on the past decade. The emphasis on quality and the definition of quality in battery products is evident in battery advertising. One ad says, "nothing can top the copper top" or "when it comes to making them last longer, we never stop". The pink bunny with a drum says "they keep going and going and going." Quality in batteries is defined as long life or high discharge capacity. This is dependent upon the purity of manganese sulfate solution supplied to electrolytic cells, the operating conditions in the production cells, the characteristics of the EMD used as cathode material and actual battery construction. High quality EMD currently used to produce "dry cells" is a black powder that contains about 92% MnO_2, has an average particle size of about 44 microns and an alkaline discharge capacity to 1.0 volts of 245 to 255 milliampere hours per gram (mAh/g) or about 900 coulombs per gram.

<u>History of EMD Production</u>

Leclanche' cells, $(MnO_2/ZnCl_2\text{-}NH_4Cl/Zn)$ were initially based on natural manganese dioxide, NMD, ores. Van Arsdale and Maier (<u>1</u>) reported the preparation of EMD in 1918. They had electrolyzed a solution of manganous sulfate. They also reported increased cell discharge capacity from cells using EMD. Between 1931 and 1949 several processes were described for the production of EMD using manganese sulfate solutions, (<u>2,3,4,5</u>). The advantages for EMD in batteries were recognized (<u>6</u>).

Commercial production of EMD in the U.S. and Japan began in the 1930s. The Japanese government (Ministry of Industry and Trade) organized a committee of EMD manufacturers and battery producers in 1948 to promote the production of EMD and improved batteries using EMD (<u>7</u>). Early Japanese EMD production was based on their domestic rhodochrosite, a manganese carbonate ore. Japanese EMD producers have been active competitors in the world markets since that time.

The U.S. Army Signal Corps in 1946 took an initiative toward developing production of EMD from domestic ores (<u>8</u>). By 1950 about 1100 tons/yr of EMD were produced by three producers. Kerr-McGee Chemical Corp., formerly American Potash and Chemical Corp., formerly Western Electrochemical Company, has produced EMD commercially for more than 40 years, beginning in 1951. The first process development programs were sponsored by the Signal Corps and based on domestic ores. Initial production was at the rate of 1 ton/day. The electrolytic cells consisted of a rectangular tank with graphite anodes and cathodes supported by the edges of the tank. Continuous deposition of MnO_2 at about 10 A/ft^2 for 30 days would produce MnO_2 plate about 1 inch thick.

Kerr-McGee's EMD plant at Henderson, Nevada has been modified and expanded many times in the past 40 years. It currently has a capacity of about 16,000 tons/yr and is now being expanded to 24,000 tons/yr.

World production capacity of EMD exceeds world demand. World production capacity is roughly 190,000 tons/yr, while demand is only about 175,000 tons/yr (<u>9</u>). This oversupplied market condition promotes fierce competition among EMD producers. Top quality EMD is essential in this competitive market.

<u>EMD Quality</u>

The first use for EMD was as a partial substitute for NMD in Leclanche' cells. The cost of EMD was considerably more than that for NMD, but batteries made from all EMD had two to three times the capacity of batteries made from all NMD (<u>9</u>).

E. Archenbach patented an alkaline MnO_2 "dry cell" $(MnO_2/KOH/Zn)$ in 1912 (<u>10</u>). The first commercial alkaline cell was pioneered by W. A. Herbert. His patents and papers published in the mid-1950s describe the early commercial cells. Alkaline cells have many advantages over Leclanche' cells. Alkaline cells have higher capacity per unit volume, are capable of higher current drain rates and are much less likely to leak than Leclanche' cells. When used in heavy-drain applications, alkaline cells can give up to ten times the service of Leclanche' cells (11). These advantages for alkaline cells have caused very rapid market growth rates and rapidly increasing demand for EMD. Battery quality - EMD quality competition became fierce.

Alkaline cells use EMD to achieve high voltages, low polarizations and maximum discharge capacity. Discharge capacity of alkaline cells has been dependent upon the quality of EMD available. Some impurities in EMD are materials that have no battery activity. Heavy metal impurities in EMD, particularly those metals with low hydrogen overvoltage, promote zinc anode corrosion and hydrogen gas evolution at the zinc anode. In the past mercury (0.4 to 0.8%) was added to the zinc to increase the hydrogen overvoltage and minimize or eliminate hydrogen gas evolution. This was necessary so that cells could be sealed with no provisions for venting. Environmental concerns were raised on this use of mercury. Most recently the mercury content of alkaline cells has been decreased and eliminated by some battery manufacturers. Consequently this has created greater demands for high purity EMD. The continuous demand for better, higher quality EMD exists to the present time.

Kerr-McGee EMD quality in 1982 was inferior to the best available. Graphite anodes were still used. Titanium anodes were being installed by some producers of EMD and the quality was improved.   A decision had to be made. Kerr-McGee either had to improve its technology and install titanium anodes or get out of the business. The decision was to stay in the EMD business and produce top quality EMD. The challenge then was how best to rapidly change production conditions to produce the highest quality EMD.

The answer was to form what might now be called a quality action team. A team was assembled that cut across functional lines. Representatives were from research and development, an engineering group, manufacturing management, quality control, mechanical engineering design, and EMD plant operations. Close communications with our customers were essential during the project and continue to the present. Problems were analyzed and divided into logical segments, i.e., cell feed preparation, titanium anode design, cell design improvement, EMD plate processing, etc. Smaller groups were assigned to address the various segments. It was important that all segments progress as fast as practical. Project team meetings were held frequently to assure that all efforts were coordinated.   Much of the experimental work was done in the production plant, based on laboratory studies at the Kerr-McGee Technical Center. Production cells were used as pilot plant cells. As improvements were made they were immediately transferred to the full plant. As an example, when it was apparent that seed crystals of jarosite promoted the removal of potassium as jarosite, solids containing jarosite were recycled to the most acidic end of the leach plant in a few days.

The team was very successful. In about one year Kerr-McGee EMD produced on graphite anodes was among the best quality for graphite anode EMD. By 1985 a patented titanium anode (12) had been developed and top quality titanium-anode EMD was being produced. This titanium anode is illustrated in Figure 1. Kerr-McGee product was declared the best in the world in 1985 by one customer, just three years after the project began.

DISCUSSION

To facilitate Kerr-McGee's study of EMD quality, a reliable method for measuring EMD performance was required. A method consisting of building cells and discharging them at known controlled rates was developed (13, 14). Alkaline discharge capacities in milliamp hours per gram or coulombs per gram could be determined accurately and reliably. Figure 2 shows the plastic cells used in Kerr-McGee laboratories to measure alkaline discharge capacity. Once Kerr-McGee's method was fully developed, numerous relationships between electrolytic cell operating conditions and EMD discharge capacity were established, based on the operation of laboratory cells by the R&D group. Since a wide range of experiments were performed with at least three variables involved, statistical methods were used to analyze the data and develop the relationships. Subsequent experience in plant cells has demonstrated that the data base generated at the R&D facility very accurately predicted actual electrolytic cell performance under a wide range of conditions. This data base is still used today to carry out a persistent, steady product improvement that does not disrupt customer processes.

To achieve the purity required for zero-mercury batteries, highly purified manganese sulfate solution must be provided to the electrolytic cells. Analytical methods were developed to measure metals in part-per-billion levels in this manganese sulfate solution. The purification process was optimized, and extra steps were taken to ensure that process upsets in the cell feed preparation portion of the plant would never result in impure manganese sulfate reaching the cells. Again, information developed by R&D was immediately applied in the plant. Since the relatively long plating cycle of EMD cells means that even short periods of out-of-specification cell feed can result in a large quantity of contaminated EMD, day-to-day consistency of the quality of that cell feed was essential. Process control techniques, such as measuring clarity with a nephelometer and precise control of sulfiding pH, were developed and utilized.

The final processing of EMD plate into product can have adverse effects on EMD quality and discharge capacity. A relative narrow particle size distribution and complete neutralization of the sulfuric acid contained in the EMD plate are important. Kerr-McGee, since the inception of its operation, had used a wet milling process, using aluminum oxide as the grinding medium. This was out-of-step with other manufacturers at that time who were using dry grinding methods. Kerr-McGee through many years of work had optimized the wet grinding procedure so as to achieve particle size distributions very similar to those achieved in the typical dry grinding system. The decision was made at least tentatively to stick with the wet grinding; and this proved to be fortuitous, as aluminum oxide is inert in a battery system, whereas wear material from typical dry grinding surfaces can contaminate the EMD with molybdenum, vanadium and chromium -- all of which greatly affect the potential for gassing in an alkaline cell.

Quality improvements achieved by the team are illustrated in Figures 3 through 6. Alkaline discharge capacity increased from about 210 to 250 mAh/g. Potassium content in EMD decreased from about 4,000 ppm to 300 ppm. Patents (15, 16, 17) were obtained on precipitating the potassium as jarosite [$KFe_3(SO_4)_2(OH)_6$] and/or alunite [$KAl_3(SO_4)_2(OH)_6$] in the leach plant. Iron content decreased from about 130 ppm to 80 ppm and became less variable. Molybdenum contamination decreased from 2.5 ppm to 0.3 ppm. Note the rapid improvements in the first few years but also the lower but continuing gains in subsequent years.

The quality of EMD currently available is characterized in Table I; however, just meeting the specification is not sufficient. The typical alkaline battery producer has production lines producing several hundred batteries per minute. The cathode mixture which is about 82% EMD is formed into an annular cathode either by impact extrusion in the can or by the compression molding of precise pellets which are recompacted against the can. To operate either of these processes, the properties of the EMD must be very consistent day in and day out. Inconsistent product would result in imperfect batteries and a high scrap rate for the battery manufacturer. Therefore, day-to-day and month-to-month consistency of product quality is just as important as the absolute level of quality. The properties of EMD that determine its processibility in a battery plant are essentially determined in the electrolytic cells with some marginal impacts created in the final processing of the EMD. To produce a consistent product, the electrolytic cells must be operated consistently. Kerr-McGee, for example, has 120 such cells in operation. Each of these cells must be operated within a very narrow range of process conditions. In addition, Kerr-McGee uses titanium anodes and copper cathodes (12, 18, 19, Fig. 1). Both of these electrodes are attacked by the cell electrolyte if not protected electrolytically. In excess of 6,000 electrical contacts have to have a reliability of at least five 9's. Kerr-McGee operating personnel, with some assistance from R&D people, developed the techniques to insure that each cell operated within the very narrow prescribed ranges and that good electrical contact was maintained under conditions not conducive to such. The inclusion of operating personnel in the Quality Action Team to a great degree motivated them to do what only they could do--operate the plant in the consistent manner necessary.

CONCLUSIONS

The effort to improve quality at Kerr-McGee Chemical Corporation demonstrated that properly organized quality action teams that cut across organization lines, that rapidly move developments from R&D to plant experimentation to full scale plant operation, and that create a common goal among the wide variety of people involved can be an essential vehicle toward achieving rapid improvement in product quality. The team created a sense of trust and cooperation which could not be achieved otherwise. Concepts were moved from R&D to production faster than reports could be written, because team members rapidly developed faith in the expertise of other team members. The authors of this paper were but a few of the members of this team which ranged from PhDs to vice-presidents to chemical plant operators, and each contributed in his own way.

The quality improvement achieved also could not have been accomplished without the development of techniques necessary to control the operation in a highly consistent fashion. The further development of these techniques continues today since, in the minds of Kerr-McGee's customers, day-to-day and month-to-month consistency of production quality is at least as important as the general level of quality. Many of these control techniques were developed by the chemical operators themselves, as they recognized and took steps to prevent the future occurrences of out-of-control operation. Thus, the dedication to improve quality continues up to today and must continue into the future in order to keep the alkaline battery in its status as the premier portable power supply worldwide.

REFERENCES

1.   G.D. Van Arsdale and C.B. Maier, Trans. Electrochem. Soc. 33, **109** (1918).

2.   C.W. Nichols, Trans. Electrochem. Soc. 57, **393** (1932).

3.   O.W. Storey, E. Steinhoff and E.R. Hoff, Trans. Electrochem. Soc. 86 **337** (1944).

4.   G.A. Lee, J. Electrochem. Soc. 95 **2** (1949).

5.   K. Takahashi, Denki Kagahu 6, **227** (1938).

6.   H. Inoue and S. Haga, Tokyo Kagyo Shikenyo Hokoku, 26, **1** (1931).

7.   A. Kozawa, in K.V. Kordesch, ed., "Batteries" Vol. 1, "Manganese Dioxide", Marcel Dekker, Inc., New York, 1974, p 436.

8.   J.C. Schumacher and R.W. Hoffman, "Electrolytic Production of Battery Active Manganese Dioxide, Pilot Run Industrial Preparedness Study", for Industrial Mobilization Division, Signal Corps Procurement Agency, Philadelphia, PA, Supply Contract No. DA-36-039-SC-716 (1951).

9.   Kerr-McGee Chemical Corporation marketing projections.

10. E. Achenbach, German Patent 261,319 (1912).

11. K.V. Kordesch, ed. "Batteries", Vol 1, "Manganese Dioxide", Marcel Dekker, Inc., New York, 1974, p 279.

12. D.A. Schulke and E.M. Spore, U.S. Patent 4,606,804, Dec. 18, 1984 (Kerr-McGee Chemical Corporation).

13. S.F. Burkhardt, "Progress in Battery Materials", ITE-JEC Press Inc., Brunswick, OH, Vol. 11, pp 136-142 (1992).

14. S.F. Burkhardt, "Alkaline Discharge Testing of Manganese Dioxide Samples" in Handbook of Manganese Dioxide-Battery Grade, edited by D. Glover, B. Schumm, Jr., and A. Kozawa, IBA Inc., pp 217-236 (1989).

15. W.C. Laughlin, V.J. Barczak, P.D. Bowerman and T.A. Rado, U.S. Patent 4,483,828, Nov. 20, 1984 (Kerr-McGee Chemical Corporation).

16. W.J. Robertson, R.C. Shaw, U.S. Patent 4,485,073, Nov. 27, 1984 (Kerr-McGee Chemical Corporation).

17. P.D. Bowerman, T.W. Clapper and W.C. Laughlin, U.S. Patent 4,489,043, Dec. 18, 1984, (Kerr-McGee Chemical Corporation).

18. O.L. Riggs, Jr., U.S. Patent 4,744,878, May 17, 1988 (Kerr-McGee Chemical Corporation).

19. O. L. Riggs, Jr., U.S. Patent 4,477,320, Oct. 16, 1984 (Kerr-McGee Chemical Corporation).

TABLE I

CHARACTERIZATION OF HIGH QUALITY EMD

Item Chemical Composition	Range
1. Available Oxygen as MnO_2	92.0 ± 0.5 %
2. Total Mn as Mn	60.0 - 61.2 %
3. Absorbed Moisture, % H_2O	1.5 ± 0.2 %
4. Fe soluble in HCl	50 to 100 ppm
5. Total HCl insoluble	0.03 to 0.10 %
6. Total sulfate as SO_4	1.05 to 1.25 %
7. Total carbon	200 to 600 ppm
8. Sodium	1500 to 3000 ppm
9. Potassium	200 to 500 ppm
10. Calcium	300 to 600 ppm
11. Magnesium	180 to 220 ppm
12. Lead	1 to 5 ppm
13. Nickel	3 to 5 ppm
14. Cobalt	4 to 7 ppm
15. Arsenic	< 0.1 ppm
16. Antimony	< 0.1 ppm
17. Copper	1 to 2 ppm
18. Vanadium	< 0.1 ppm
19. Titanium	0 to 30 ppm
20. Chromium	0.4 to 2 ppm
21. Molybdenum	0.2 to 0.5 ppm
22. Aluminum	50 to 300 ppm
Physical Properties	
23. pH	6 to 7
24. Compressed density	3.05 to 3.13 g/cc
25. Apparent density	25 to 27.4 g/in^3
26. -200 mesh	86 to 93 %
27. -325 mesh	60 to 72 %
28. Initial open-circuit voltage	1.610 to 1.630 volts
29. Alkaline discharge capacity	245 to 255 mAh/g
30. Surface area	28 to 35 m^2/g

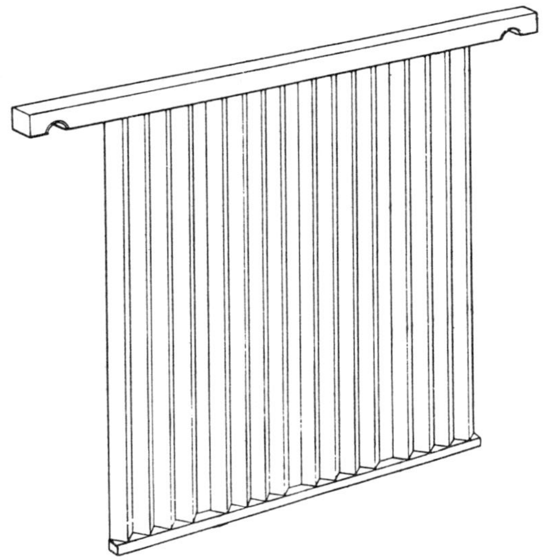

Fig. 1 Kerr-McGee Patented Anode

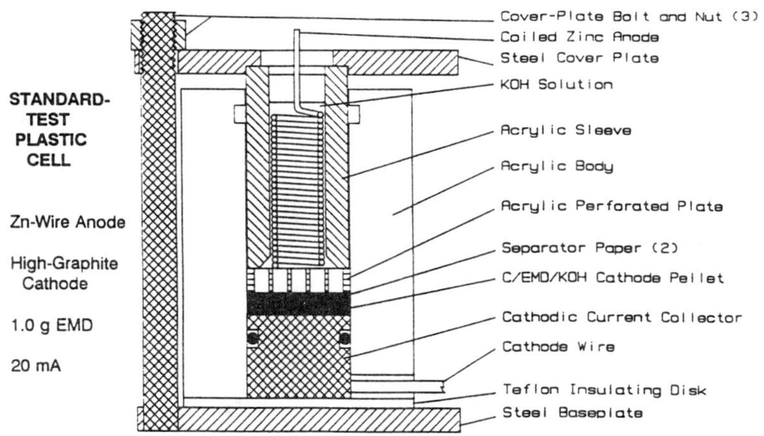

Fig. 2 Cross Section of Reusable Discharge Cell

57

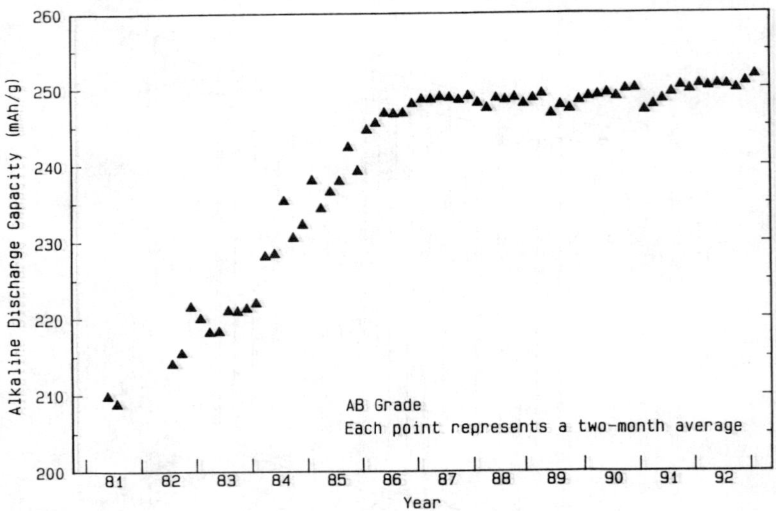

Fig. 3 Discharge Improvement of Kerr-McGee EMD

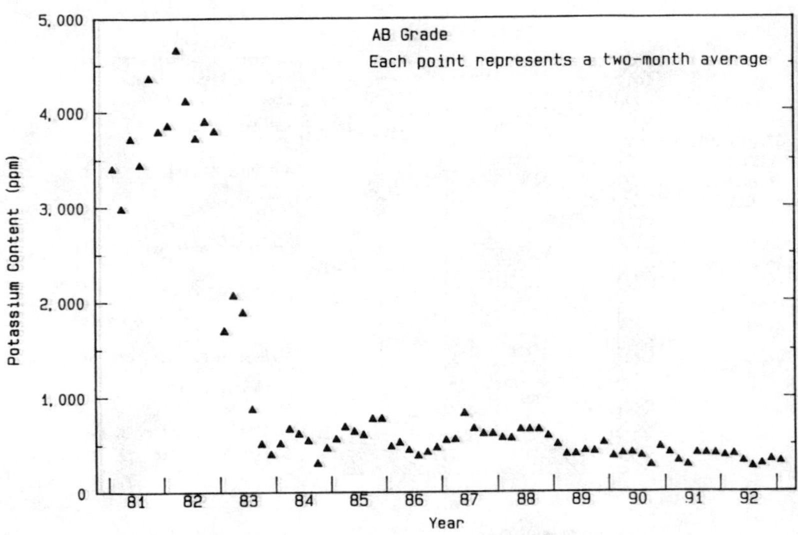

Fig. 4 Improvement in Potassium Content of EMD

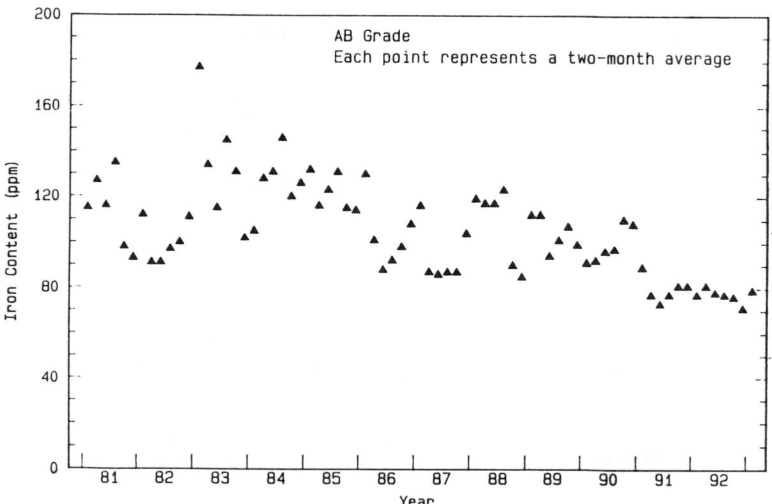

Fig. 5 Improvement in Iron Content of EMD

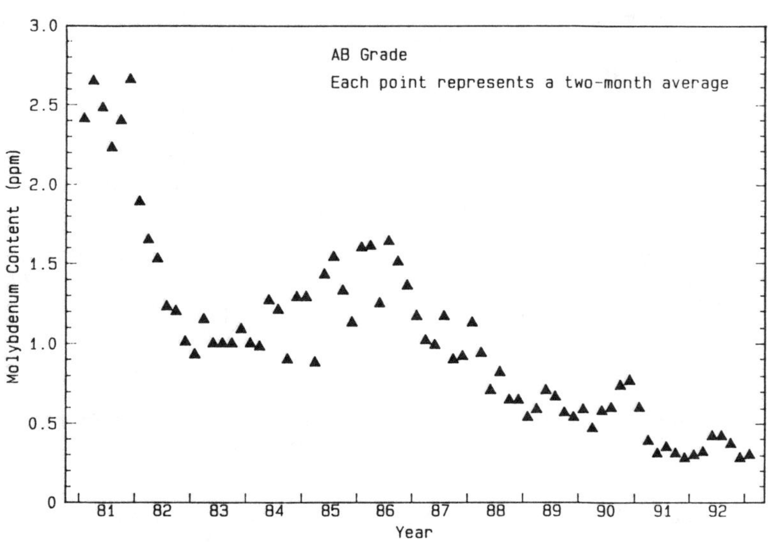

Fig. 6 Improvement in Molybdenum Content of EMD

QUALITY MANAGEMENT ACTIVITIES TARGETING "QUALITY FITNESS NO. 1"

Azumi Inaba
Production & Technical Control Division, Oita Works,
Nippon Steel Corporation
1 Nishinosu, Oaza, Oita 870, Japan

Waves of diversification and increasing strictness of customer requirements for product are natural consequences also in the steel industry which is a material industry.

Based on this recognition the Oita Works of Nippon Steel conducts quality management activities on the "Customers First" principle for actions and targeting "Quality Fitness No. 1".

This paper describes the background which led to these activities and the elements which support them.

INTRODUCTION

Oita Works reached last year the 20th anniversary of the inauguration of the integrated iron and steel production structure with the blow-in of the No. 1 Blast Furnace in April 1972.

As the first steelworks after the establishment of Nippon Steel, Oita Works was constructed incorporating pioneer technologies of those days in order meet increasing steel demand resulting from Japan's high economic growth. Determined efforts to develop future technologies, instead of using only completed and proved technologies, realized numerous innovative processes including the epoch-making fully continuous casting process and fully computerized systems and have continued to serve as the sources of this works capabilities up to the present. The Japanese steel industry was confronted with severe business slumps due to the two oil crises and the unprecedented appreciation of the yen. However, Oita Works tided over these difficulties with technological innovations and measures to reform its basic makeup, and established on 8 million ton crude steel production structure with the startup of the No. 2 Blast Furnace in 1976. At present Oita Works meets internal and external requirements exactly and rapidly as a steelworks producing mainly hot-rolled coils and plates and boasting the world's top level quality and productivity.

DEVELOPMENT OF OITA WORKS AND TRANSITION OF QUALITY MANAGEMENT

The history of the development of Oita Works is the history of the development of quality management. The purpose of quality management is, of course, to manufacture products meeting customer requirements efficiently and at low prices, deliver them to customers without delay and assure their quality positively. To achieve this purpose we continued our strenuous efforts and studies. In the varying work environments, quality management itself attained growth and development. At present Oita Works conducts quality management activities on the "Customers First" principle for action and targeting "Quality Fitness No. 1".

(1) Quality management of "emphasis on inspection" which opened up the fully continuous casting process age

During the "genesis" period of Oita Works from 1972 to 1975, the works which started fully continuous casting for the first time in the world endeavored to establish the basis of continuously cast products. Japanese continuous casting technology at that time was at its dawn. Continuously cast products accounted for only five percent of Japan's total crude steel production, and the main products were manufactured from ingot mold steel by the ingot making to primary rolling process. At such a time the exclusive introduction of a new production system based on fully continuous casting meant staking the works fate, and we took great pains in establishing product quality required by customers.

Although continuously cast products had excellent quality characteristics, the replacement of ingot products which customers were accustomed to use by continuously cast products required more strict quality assurance than before.

Therefore, the most important item of quality management in the early days was to fit the quality level of continuously cast products for use by thoroughly rotating the PDCA cycle for the evaluation and improvement of the quality of these products. We made not only quality evaluation of large quantities of products at the works but also quality evaluation of processes at customers plants. Problems posed were studied by close cooperation between a staff of experts and people concerned at the works, and measures to solve them were taken immediately. The MR (mill representative) system which our works adopted for the first time in Japan effectively functioned in obtaining cooperation from customers and served for the timely solution of problems.

Although the quality of continuously cast products was remarkably improved in this way, we emphasized inprocess inspection and final inspection before shipment in order to prevent the continuous occurrence of large-quantity quality defects and increases customers reliance on our products.

In 1975, Nippon Steel received the Japan Quality Control Prize for the first time in the history of the steel industry for the execution and distinguished results of quality management.

(2) Quality management of "emphasis on integrated process quality assurance" which opened up the direct and directly-coupled process age

In October 1976, the No. 2 Blast Furnace was blown in, and our long cherished desire to start twin-furnace operations was accomplished. However, the oil crisis faced the Japanese economy, and the economic situation changed from high growth to low growth. Nippon Steel enforced various policies in quick succession in order to "make itself into a company possessing a basic makeup and earning power well capable of maintaining profitability even with 70 percent-capacity operations". Important items of quality management were promoted in parallel with cost reduction, energy saving, productivity improvement, etc. The "Striving Oita, C & Q" slogan of Oita Works in 1978 emphasized the eagerness to strive after the limits of cost reduction and make quality evaluation indisputable by the united efforts of all the people at the works.

What was realized against such a background was the "Directly-Coupled Process-v" in which the steelmaking process and the rolling process were directly coupled by continuously cast slabs of high temperature. Continuously cast slabs from the steelmaking process were conventionally cooled to normal temperature once, carefully inspected, conditioned and transferred to the rolling process, in which they were reheated and rolled into products (hot-rolled coils and plates). The important point of the directly-coupled process-v was the elimination of the need for slab inspection and conditioning by integrated process quality assurance which was made possible by standardizing and making patterns of the operating conditions in the steelmaking process.

The new quality management of "emphasis on integrated process quality assurance" which superseded the old one of "emphasis on inspection" and which opened up the age of direct and directly-coupled processes contributed a great deal to cost reduction and energy saving, in addition to quality level improvement.

(3) Quality management of "emphasis on Source Management" which opened up the "Market In" age

The Japanese steel industry suffered a direct hit from the worldwide depression and was placed under such a severe management environment that crude steel production dropped below 100 million tons in both 1982 and 1983. Under the influence of the change in demand structure caused by the recessions after the two oil crises, market and customer requirements for steel products became diversified and more strict year after year. Up to that time the works had a strong tendency to respond to quantitative requirements and cost reduction requirements making use of its production structural features suitable for mass production. However, under the works medium-term plan started in 1983, we began to make production efforts clarifying the change "from quantity to quality" and the change in our awareness and actions "from Production Out to Market In".

In order to meet customer requirements, we strengthened the sources of quality management,such as market research, product planning, product development and quality and process design, and systematically and strategically promoted measures to strengthen quality-equipping power, measures to develop new products and measures to improve process capability As a result, process capabilities such as dimensional accuracy and surface quality made rapid progress while on the other hand many new products such as hot-rolled coils of high strength, high ductility and excellent formability and plates of high strength, high toughness and excellent weldability were developed to meet customer requirements.

Later on quality-related organizational reinforcement was made by newly establishing the Quality Assurance & Planning Department and the post of general manager in charge of quality, and "Quality Management Rules" were instituted and enforced to define basic ideas of quality, basic mechanisms of quality management, the sharing of responsibilities among organizations and procedures for quality management. Consequently, the quality management basis at the works was steadily improved, and the evaluation "Oita of Quality" became established.

(4) Quality management of "Customers First" which opened up the CS (customer satisfaction) strategic age

Under the new medium-term management plan formulated in 1991, Oita Works was clearly defined as a steel supply base for the company. This fact shows that Oita Works is charged with the mission to stably supply large quantities of steel having outstanding quality and cost competitive-ness so that Nippon Steel continues to secure target earnings for the steel business.

Based on the company's new medium-term management plan, the work prepared a new medium-term plan for th period from 1991 to 1993. One of the features of this plan was that the targets to be achieved in the three-year period regarding software tasks of quality management and the principle for actions shown below were newly indicated, in addition to hardware tasks to be performed mainly for the strengthening of steel cost competitive-ness and the ability to respond to the market requirements.

The "Quality Fitness No. 1" which forms part of the title of this paper means that we aim to make our works a No. 1 works in respect of the degree of satisfaction it gives customers and the degree of its quality competitiveness. We intend to live up to this principle by manufacturing products of optimum design quality (target quality) and manufacturing quality (actual quality) with fitness (easily, regularly and economically) and delivering products fit to fulfil the expectations requirements of customers.

We dared to select this unoriginal motto "Customers First" as our principle for actions because we wanted to put into it our eagerness to do common things steadily and in a common way from the standpoint of "customers satisfaction" and "actual practice" from which quality management should start.

More accurate quantitative evaluation yardsticks than in the past are needed to promote "Quality Fitness No. 1". Although CSA (customer satisfaction audit) is now under trial for the evaluation of the degree of satisfaction of customers, we are still groping for a yardstick regarding the fitness of manufacture.

Targets of the New Medium-Term Quality Management Activities of Oita Works.
Principle of activities : "Customers First"
Target to be achieved : "Quality Fitness No. 1"
1. A steelworks which has comprehensive ability to quickly meet customer requirements
(1) The works should have process capability to quickly meet customer requirements.
2. A steelworks which can supply new products timely to the market.
(1) The works should be able to correctly grasp customer requirements.
(2) The works should have basic makeup (planing ability) to connect market needs and wants to new product development and commercialization.
(3) The works should have wisdom and a management structure sufficient for new product development.

---

3. A steelworks which rotates the PDCA cycle on standardization and spirals
  up quality management.
(1) The works should certainly accumulate and transfer technology.
(2) The works should clarify key points of quality management.
(3) The works should clarify rules for carrying out quality management jobs.
4. A steelworks where everyone has a firm belief in the significance of quality
  management.
(1) The works should secure basic competence for quality management and
  improvement, early of abnormalities, etc.

I have so far briefly described basic ideas, background and feature of today's quality management, tracing the 20-year history of quality management at Oita Works. Quality management is called a science of practice, Therefore, I would like to show below two typical examples in detail.

EXAMPLE 1 - "PDCA FOR TARGET ACHIEVEMENT"

(1) Relationship between medium-term quality targets and annual quality
    targets

Oita Works is now devoting all its energies to the achievement of the targets of the medium-term plan for the 1991-1993 period. The medium-term quality targets are part of these targets and at the same time the bases setting the targets of all activities for the improvement of cost, personnel, material flow, etc. This is because of the works basic idea regarding quality management that the manufacture of products capable of attaining customers satisfaction takes priority.

The medium-term quality targets consist of hardware-type tasks such as individual product quality improvement targets and new product development targets and software-type tasks such as quality management structural improvement targets.

The former tasks are called quality plans, and they consist of "product plans" which are prepared regarding measures to commercialize products to be planned, developed and marketed by the works, "quality target plans" which are prepared regarding the targets of the quality characteristics of the works products to be achieved and measures to achieve the targets and "quality assurance plans" which are prepared regarding tasks for quality assurance such as the completion of various kinds of sensors for quality assurance and measures to carry out such tasks.

In the steel industry, it takes at least one to two years or three or more years in some cases to introduce new equipment or to develop a new technique or product. Therefore, although targets to be achieved in a medium term are set for two or three years hence, there are also annual targets necessary for approaching them, and the annual targets, together with the measures to achieve them, are broken down on a very small scale to from each department's execution plans.

It is necessary to set annual quality targets first and then grasp medium-term quality targets from the aspect of the execution of the annual quality targets, This procedure serves to clarify the assumptions and conditions on which improvement strategies should be promoted with foresight and taking a broad view. Therefore, annual and medium-term quality targets cannot be considered separately.

When the management environment changes so rapidly as at present, we are often compelled to alter our targets for the next year, not to speak of those for two or three years later. Therefore, plans must be flexible, and we need to study what have changed and how and why they have changed so that we can correct our course at any time and also need to have second and third targets and measures.

(2) Flow of the establishment of medium-term quality targets

In order to set quality targets, it is necessary to clarify when and what quantities and quality levels of products are needed for what customers. In addition, it is necessary to analyze factors based on which targets must be set. such as the background of market needs, comparisons with competitive products and manufacturers and influences of the failure to achieve the targets.

The works quality targets are primarily established based on the company's medium-term management plan, but they are judged from market trends and technology trends classified by product series and demand area from the medium and long-term viewpoint based on the works own market research results. For this reason, the common possession of customer information by MR activities which will be explained in detail in "Example 2" has an important meaning.

As quality targets serve as the bases for preparing the works relevant plants, original draft targets are prepared a few months in advance. After being discussed in the various specialty fields, the original draft targets are discussed and decided by the "Medium-Term Job Council" headed by the General Superintendent of the Works and are then distributed to all relevant departments of the works in writing. Quality targets are reviewed from to time through fixed steps, and relevant plans are revised simultaneously.

(3) Flow of the establishment of annual quality targets

In setting annual quality targets, it is first necessary to establish quality policies for the year which serve as guiding principles for them. The quality policies are based on the works management policies for the year indicated in the General Superintendent's address to all the people of the works at the beginning of the year and are drafted taking into account the results of evaluation of the achievement of quality targets and changes in the management environment and through discussions among relevant departments and among relevant hierarchical posts. The original draft policies are discussed by the "Product Quality Conference" headed by the Deputy General Superintendent and finally decided by the "Management Conference" headed by the General Superintendent.

Since annual quality targets have the character of interim management index for medium-term quality targets, their establishment requires concrete management items and indices which are connected to various measures to be taken in the year by each department. If, for instance, large-scale equipment investment in necessary for the achievement of medium

-term quality targets, the test, study and budget preparation needed to realize the equipment installation constitute part of the quality targets for the first year of the term.

(4) Execution and follow-up of quality targets

The basis of the method of quality management at Oita Works is to spiral up quality management by taking each of the following steps P, D, C, A without fail.

> (P) Establishment of quality targets and preparation of quality execution plans to achieve the targets
> (D) Certain execution of the quality execution plans
> (C) Check of execution results and execution processes and mechanisms
> (A) Revision of the targets and execution plans or improvement of the execution processes and mechanisms

Among these steps, check and action are regarded as most important at Oita Works, Therefore, when annual quality execution plans are to be prepared, the quality staff prepare quality follow-up plans in advance.

Quality targets are followed up at the committee meeting every month, but the priority of discussion at the meeting is placed more reflection on the execution processes and mechanisms which have led to the target achievement results and the methods to be used to improve the processes and mechanisms based on the reflection than on the target achievement results themselves. Correction of course is added to the annual quality targets and execution plans prepared the beginning of the year at the meeting, if necessary.

Another important thing at the follow-up stage is to take precautions against deflection from the course to be followed by the works as a whole or postponement of more important tasks because of the occurrence of various problems in routine work. To this end also, follow-up plans should be pursued without fail.

(5) Quality-related assembly bodies

Medium-term and annual quality targets and annual quality execution plans are prepared based on the works basic ideas and management policies regarding quality and through fixed steps. If the condition based on which such targets and plans are prepared change, such targets and plans are changed suitably and coordinated with other relevant plans.

In order to secure both rigidity and flexibility of plans and in order also to promote mutual understanding and the common possession of information among hierarchical posts and among departments, the works have several assembly bodies.

The works supreme decision-making body is the Management Conference headed by the General Superintendent. This conference meets once every week to discuss all daily important items including quality problems. The works supreme body to discuss quality is the "Product Quality Conference" headed by the Deputy General Superintendent-Technical. This conference meets once every month to prepare quality policies and quality execution plans, discuss the correction of course and follow up the condition of their execution. The General Manager - Quality is responsible and authorized to manage the "Product Quality Conference".

The "Product Development Study Team " headed also by the General Manager - Quality meets regularly once every month to discuss practical matters concerning product development and marketing.

In addition to the above, various other assembly bodies involving different organizational posts both vertical and horizontal are functioning. They range from the "QM (Quality Meeting)" at which those in charge of practical business get together from the manufacturing, quality control process control areas to discuss daily quality problems regularly once every week to the "Morning Meeting" held every morning at each department (See Figure 1).

EXAMPLE 2 - "CONSOLIDATION OF THE BASES OF ACTIVITIES"

Regarding the "quality function" and "men" which are the bases of quality management activities, Oita Works puts particular emphasis on the common possession of quality information, especially customer information.

I would like to explain below the actual conditions of the roles and activities of the "MR (Mill Representative) Activities" which are the key point of the "common possession of customer information ".

(1) Roles of MR activities

The greatest feature of the MR activities is that veteran engineers belonging to the organization of a steelworks and well versed in the actual situations and manufacturing technologies of the steelworks go to customers in person, grasp the functions and characteristics of products required by the customers and reflect them timely on the decision of manufacturing conditions, quality improvement, new product development, etc., at the steelworks.

Oita Works was the first to adopt the concept of the MR activities in Japan. At the outset of the start of operation of our works when the continuous casting ratio of Japanese steel products was still below five percent, it was unimaginably difficult for us to have our continuously cast products used by customers without concern. Almost three years were needed for our engineers in charge in those days obtain customers full reliance on our continuously cast products after visiting them innumerable times.

Later on the continuous casting ratio remarkably increased, and the MR activities began to have the two important roles of direct technical service and support activities for customers and market research intended for marketing strategies, instead of the initial of analytical approach to quality management by reflecting the results of use by customers on our next manufacture (See Figure 2).

To the job positions of MR, therefore, we assigned excellent engineers well informed of steel manufacturing technologies such as superintendents in the manufacturing sector and experts in design and quality control and strengthened the staff in charge.

(2) Actual conditions of MR activities

The PDCA cycle for the MR activities is rotated during the period three months earlier than the period during which that for the quality management activities is rotated (i,e., from January to December of the year). This is because the customer trends and market technical trends observed as a result of the MR activities are reflected on new product and technology development plans and equipment plans and not on the establishment and renewal of annual quality targets.

At the planning stage of the MR activities, the segmentalization of markets and customers according to some intended standards and the argument of technical strategies corresponding to the segmentalization are the key points. For instance, when how our works technical seeds should be connected with customers latent needs is a problem, what products in what processes of what customers our approach should be aimed at is argued and decided.

At the execution stage, it is very necessary to study the following items for market research and carry out the following items for customer service both systematically and according to plans.
- Examples of items for market research -
    (a) Changes in demand structure
    (b) Changes in supply structure
    (c) Changes in environment and way of using products, and in required function, performance, reliability and degree of satisfaction
    (d) Trends in quality and technical ability
    (e) Trends in auxiliary technologies and change in, and influences of, social requirements such as politics, economics and regulations
- Examples of items for customers services -
    (a) Grasp of the quality evaluation of product shipped and reflection of the quality evaluation on design and manufacture
    (b) Technical services to customers regarding the use of products and support to customers in solving problems
    (c) Evaluation of the usability of trial products at customers
    (d) Activities for spreading new products among customers
    (e) Settlement of claims and prevention of recurrences of trouble

Each time the results of MR activities are obtained they are compiled into a report as "market technical information". The report is submitted to people concerned at the head office and at the works. All information given in the report is input to an optical disc, and at the same time an abstract of the information is registered in the company's general technical information system. This internal information network system allows the common possession of all market technical information.

Comprehensive summarization, analysis, evaluation and prediction of MR information and study of policies to solve problems are conducted semiannually. After being discussed at meetings of the "Product Quality conference", the study results are reflected on new product and technology development, marketing strategies, equipment plans, etc. At these meetings, the mechanisms and procedures of MR activities are checked and necessary actions are taken in order to raise the levels of MR activities.

CONCLUSION

To meet customer requirements and secure appropriate profits for the company, it is necessary to have a quality management system which makes it possible to continuously develop and manufacture products fit for the purposes of both customers and the company, while fulfilling the requirements of the social environment.

The following are the main points of the two examples of quality management activities at Oita Works.

(1) The establishment of a structure which allows "rigid" execution of activity plans to achieve quality management targets and "flexible" alterations in plans according to the change in actual conditions.

(2) The promotion of systematic and strategic gathering of customer information and common possession of the information gathered. The establishment of parallel job positions capable of proposing quality improvement and product development based on the information.

With the 21st century just around the corner, Oita Works always keeps abreast with the times feeling keenly the dynamic movement of the world. The works is promoting the restructuring of its management basis aiming at the creation of the world's top-class steelworks capable of realizing dreams of future technologies. Also in the aspect of quality management, the works is devoting all its energies to the building of a new quality management system combining the ISO 9000 series with the Japanese TQM (total quality management).

Lastly, I hope that this paper will contribute to some extent to the improvement of your quality management activities.

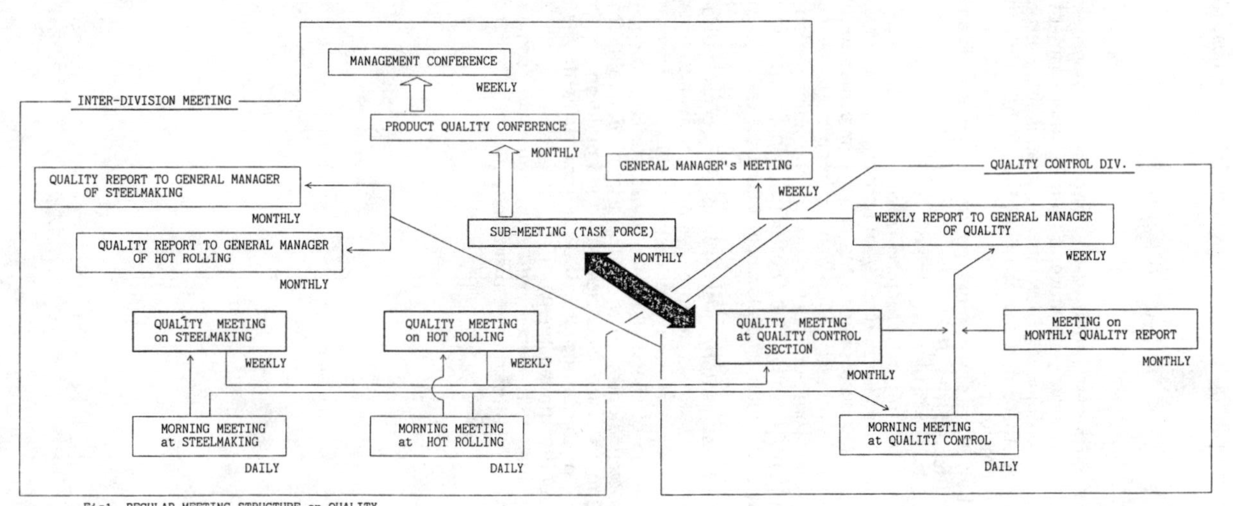

Fig1. REGULAR MEETING STRUCTURE on QUALITY

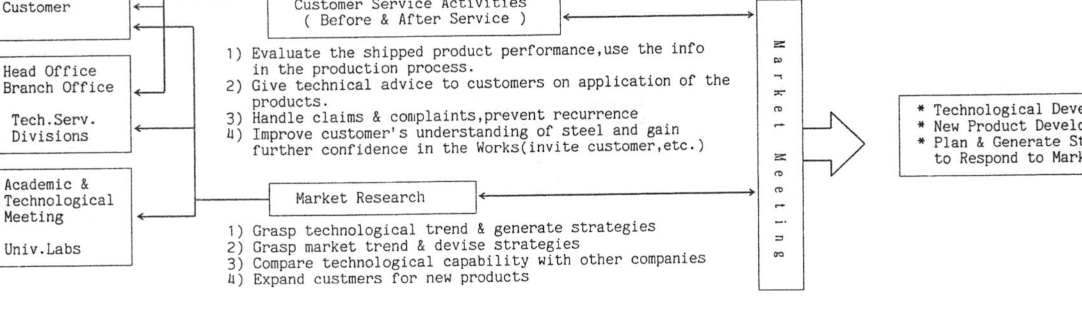

Fig2. MR Activities

ELEMENTS OF A QUALITY SYSTEM

Leslie W. Flott, C.Q.E., ASQC Fellow
Manager, Customer Satisfaction
Wayne Metal Protection Company
Fort Wayne, Indiana

Author, "Quality Control" column,
Metal Finishing Magazine

Abstract

Ensuring customer satisfaction isn't the beginning of a quality system but the end result of a quality system. The elements of a quality system include:

    1 : Designing it right
    2 : Ordering it right
    3 : Building it right
    4 : Shipping it right
    5 : Doing it all on schedule.

A successful program consists of doing what is necessary, when necessary, the best way we know how. Effective Quality Systems establish guidelines for the consistency and quality of output for every aspect of an operation.

INTRODUCTION

Every successful company places the quality of the products or services that it produces high on its list of priorities. They provide products or services that satisfy specific needs, which meet or exceed customer requirements, while complying with applicable standards and specifications, remaining competitive and still yield a reasonable profit. The achievement of these goals means being organized to reduce, eliminate or prevent defects.

A company attains these goals through the development, implementation and maintenance of systematic quality initiatives, i.e. a Quality Management System (QMS). ANSI/ASQC Q94-1987 (ISO 9004) provides an excellent guideline for such a system.

A vendor can delight its customer with a product or service by:

     1 : Designing it right
     2 : Ordering it right
     3 : Building it right
     4 : Shipping it right
     5 : Doing it all on schedule.

In general terms, a successful program consists of doing what is necessary, when necessary, the best way we know how. An effective QMS establishes guidelines for the consistency and quality of output for every aspect of an operation.

A winning QMS must address both the company's and the customer's needs. These are usually more closely related than is commonly believed. A company needs to deliver to its customer the best products at the best prices and still make a reasonable profit. Customers must be confident that the company will deliver the required quality the first time and every time. Regardless of other considerations a QMS must provide objective evidence in the form of data that the system is producing products or services of an acceptable level.

The responsibility for and commitment to a quality policy belongs to the highest possible level of management. The QMS must like any other policy be in complete agreement with the objectives enumerated in other company policies. In order for the company's quality policy to work it must be clearly stated, easily understood by everyone, fully implemented and maintained.

DEFINING THE QUALITY SYSTEM

A quality system is the organizational structure that establishes and manages the responsibilities, procedures, control of processes and the means by which they are accomplished. A quality system based on stated company policies must also be specific to the type of business being conducted. It must be understandable, understood, effective, able to satisfy customer expectations and should place its primary emphasis on prevention rather than detection of defects.

Every aspect worthy of concern after parts are finished is worthy of concern before the parts are finished. This means that instructions must be clearly understood by the person who designs the part, the person who purchases the part, the person who finishes the part, the person who inspects the part, the person who assembles the part, and the end user. If this is done, the customer will consistently receive the services that are needed, with a minimum of problems, rather than the services that someone thought that they wanted and didn't give them.

DESIGNING IT RIGHT

Product quality begins with product design, but an engineering design is not something to simply throw over the fence to production. Designs must be reviewed by all operating functions (i.e. purchasing, marketing, manufacturing, quality, etc.) before finalizing the quality requirements of any product. There must be an absolute understanding with customers as to exactly what is expected in the finished goods. These requirements must be clearly and accurately communicated to everyone in the organization.

Product requirements must be provided to everyone in a document that everyone can understand. This document, called a Product Brief, should include all pertinent performance characteristics (i.e. what environment will the product see, for how long etc.), sensory characteristics (i.e. appearance, feel etc), tolerances (i.e. applicable standards or specifications), what assurance or verification of quality is required.

Specifications and tolerances must not be used without thought as to how they translate into dollars. It is axiomatic that tolerances cost money. It is clearly less expensive to cut logs for the fireplace to 20" +/- an inch, than it is to make wooden table legs to 20" +/- 1/64". The designer's goal must be meaningful tolerances.

<u>**ORDERING IT RIGHT**</u>

Customers' purchase orders, drawings and
specifications determine the nature of the
products or services that are expected to be
supplied. The customer, whether internal or
external, must be responsible for providing copies
of the latest revisions of their drawings and
specifications, since a vendor has no way of
knowing when the customer makes an engineering
change (EC). Sales receives EC's from the
customers and is responsible for passing these
changes on to Production. Customers' drawings
should be stored in convenient files and their
receipt acknowledged in a manner that is
satisfactory to the customer. Production is
responsible for charging out and using the latest
EC's, drawings and specifications.

All parts and materials should be
received, logged and inspected after logging in.
The receiving inspection is to verify that the
parts or materials are as described in the
accompanying paperwork. Receiving inspection
identifies all accepted lots and sends them to
stock. Rejected lots are identified and set aside
in a designated holding area for disposition. QC
follows up to see that appropriate corrective
action is taken by any customer or supplier who
has shipped defective parts or materials.
Instructions for inspection must be followed
conscientiously. QC should review the receiving
records to determine if any suppliers, or
customers, are consistently failing to meet
standards, the company's or their own.

Whenever possible, all materials and
articles should be identified by a basic part
number and revision number. Critical materials and
articles should also carry a serial or lot number.
If required, a separate list of materials and
articles by identification number should be
attached to the customer paper work or be made
part of an appropriate Certificate of Compliance.

Copies of all purchase orders and
associated documents, regardless of source, must
be kept on file for customers to review. Incoming
raw materials must be marked in accordance with 29
CFR, Part 1910, Subpart 2, Paragraph 1910.1200, or
other applicable regulations, so as to be properly

identifiable and traceable to source and should be stored apart from the normal flow of in-process material until accepted. All raw material must be certified, traceable to purchase order, date of receipt of the material and the inspection of the material. Non-conforming material must be segregated from other material until its disposition has been determined.

BUILDING IT RIGHT

Production planning is intended to ensure that all processes are operated under known and controlled conditions. These conditions include controls on raw materials, production equipment, processes, properly trained and instructed personnel, utilities and environments. Implicit to an effective control system is the feed back of information following any change to the process or equipment. Any change in the relationship between process and product possibly resulting from that change must be documented and appropriately communicated.

Production control should include a system of preventive maintenance, with special attention paid to those machine characteristics that may contribute to critical product quality characteristics.

Work instructions may consist of procedures for common practice or specific instructions for a given product, but in either case should be published and available to the operator. These documents should include both work instructions and workmanship criteria for determining what work is or is not acceptable. The use of visual aids may be appropriate in this regard.

A system for verifying the status of a product must be in place. This system may include check sheets, stamps, SPC charts, process inspection matrixes or other appropriate records. The intent of all such systems is to maximize yields while minimizing the possibilities of error. This also means that all in-process inspections must be planned and executed according to plan. In-process inspection also requires procedures for maintaining positive unit identification and control (i.e. lot control).

Before parts are released for production samples must be run using production equipment and processes, regardless of whether samples have been run on pilot equipment or not. Acceptance sampling for initial lots must demonstrate a higher level of process capability than is required by the customer for final production. This approach is based on the assumption that initial runs will always be inspected more thoroughly, including 100 % inspection, than will ongoing production. Thus a part that requires a long-term Cpk of 1.66 might be held to a Cpk of 2.00 for an initial submission. A process that is incapable of producing to customer expectations when being extraordinarily monitored is unlikely to be able to meet those expectations under less stringent monitoring.

An indispensable part of production control begins with obtaining and understanding the appropriate specifications. Control must also include monitoring the process yield through sampling or testing. Understanding and using an effective system for determining the results of what is being produced, the maintenance and control of all solutions and finally the control and maintenance of the means by which the product quality is measured is the essence of production control.

All items must be tested or sampled as required by customers' specifications. Final inspection and tests can be performed either on 100 % basis or on a sample of the items. A sampling plan may be performed according to MIL-STD-105E, or ASTM B 602.

Once regular production has begun a more or less conventional sampling program comes into play. If the customer specifies MIL-STD-105E, or some other specific sampling plan, it must be used, otherwise it might be preferable to use a form of continuous process control. Experience suggests that the longer process controls are in place the better the overall quality of the process yield becomes, benefiting both supplier and customer.

All processes should be verified as to their ability to meet product specifications. Once process control has been established and verified, adequate control must be maintained to assure the characteristics remain within appropriate guidelines. It is essential that control be continuous. One procedure for continuous monitoring is called Statistical Process Control (SPC). Unfortunately, many American workers lack the necessary mathematical skills to run SPC.

PRE-CONTROL

An alternative to continuous process control preferred by many is Pre-control, a modified Statistical Process Control (SPC). It was originally invented by Rath and Strong in the 50's and popularized by Dorian Shannin. This version, adapted by this author, doesn't require advanced mathematics. The specification range is divided into four equal parts.(see Figure 1) The area from the mid-point of the specification (MP) to plus or minus half way to the specification limits is defined as the Green Zone. The area between the Green Zone and the specification limits on both sides of the Green Zone is defined as the Yellow Zone. Any point beyond the specification limits is in the Red Zone.

```
                                    RED ZONE

plus 4 sigma    |-----------UPPER SPEC LIMIT------
                |                   YELLOW ZONE
                |
plus 2 sigma    |---------------------------------------
                |
                |                   GREEN ZONE
average         |-----------MID POINT-----------
                |                   GREEN ZONE
                |
minus 2 sigma   |---------------------------------------
                |
                |                   YELLOW ZONE
minus 4 sigma   |--------LOWER SPEC LIMIT--------
                                    RED ZONE
```

Figure 1: Control chart for Precontrol

Five (5) simple rules then allow the control of production quality, as follows:

Rule 0: If <u>zero (0)</u> points are outside of the Green Zone, except for possible adjustments called for in **Rule 4**, continue to run production.

Rule 1: Any <u>one (1)</u> data point in the Red Zone means the process is out of control, the supervisor is notified, the process is shut down and the problem corrected before any more parts are run.

Rule 2: Any <u>two (2)</u> out of <u>three (3)</u> data points in the Yellow Zone means the process average may be drifting, further sampling must be monitored closely.

Rule 3: Any <u>three (3)</u> out of <u>five (5)</u> points in the Yellow Zone means that the process is approaching an out of control condition and should be treated the same as under Rule 1.

Rule 4: If more than <u>four (4)</u> data points are on one side of the mid-point, even if still in the Green Zone, adjust the process as in Rule 2.

<u>PROCESS CAPABILITY STUDIES</u>

Process Capability is the inherent reproducibility of a product turned out by a process. In determining the "Capability" of the process, it is necessary to compare the output of the process against the tolerance of the product, i.e. to measure the design tolerances against the process output produced. It is therefore necessary, under most conditions, to use a separate chart for each part number.

In general it is also desirable and necessary to calculate a separate Process Capability for each individual part number. While capability is independent of the process, common practice is to calculate the process capability only when the process yield is normally distributed and in a state of statistical control. In other words, before attempting to calculate the capability of the process producing a specific part number, the control chart should show only a natural distribution pattern.

Once it's decided that the process meets these criteria, the capability can be determine in one of three ways:

1) the C_p, the Capability Index

2) the C_R, the Capability Ratio

3) the Cpk

Cpk is probably the most widely accepted measure of process capability in current use. While it has some limitations, it overcomes the limitations of the other methods mentioned above. Cpk is calculated by:

Cpk Worksheet

1. Upper Specification Limit = USL =

2. Lower Specification Limit = LSL =

3. Process Average = $\overline{X}$ =

4. Process Standard Deviation (Estimated) = s =

5. Calculate Cpk's:

<table>
<tr><td>Cpk Lower</td><td>Cpk Upper</td></tr>
<tr><td>$\dfrac{\overline{X} - LSL}{3\ s}$</td><td>$\dfrac{USL - \overline{X}}{3\ s}$</td></tr>
<tr><td>$\dfrac{\rule{2cm}{0.4pt} - \rule{2cm}{0.4pt}}{3\ s\ \rule{2cm}{0.4pt}}$</td><td>$\dfrac{\rule{2cm}{0.4pt} - \rule{2cm}{0.4pt}}{3\ s\ \rule{2cm}{0.4pt}}$</td></tr>
<tr><td>Cpk Lower = _______</td><td>Cpk Upper = _______</td></tr>
</table>

Cpk = the smaller of the Cpk Upper or Cpk Lower

Cpk = ____________

This Chart shows the relationship between Cpk and defects in any lot.

Cpk	Defects/Million	% Defective
0.33	317,310	31.7
0.50	133,614	13.3
0.66	45,500	4.5
0.83	12,419	1.2
1.00	2,700	0.27
1.17	465	0.046
1.33	64	0.006
1.50	6.8	0.0007
1.67	0.6	0.00006
2.00	0.0018	0.00000018

Most customers demand Cpk's of 1.33 or better. Many customers require and expect defects to be no more than a relatively few per million. Verification at these levels cannot be economically obtained by any sampling plan based on MIL-STD-105E, MIL-STD-414, Dodge-Romig tables or ANSI Z1.4 or other popular sampling schemes. This is itself a strong argument for using SPC.

One other important modification to conventional sampling plans which should be employed is to always use a zero (0) acceptance number (i.e. one (1) defect regardless of sample size will always reject the order) regardless of the plan being used or specified by the customer. Thus the Average Outgoing Quality Level (AOQL = % Defective * Probability of Acceptance) equal to or better than the customer's requirements is guaranteed at all times.

GAGE AND INSTRUMENT CALIBRATION

All gages, measuring instruments, sensors, related computer software, manufacturing jigs, fixtures and process equipment, as well as, all test equipment must be properly identified and calibrated to standards traceable to the National Institute for Standards and Technology (NIST, formerly N.B.S.). Each new gage or instrument must be inspected before being issued for use. A schedule for calibrating gages and measuring and test instruments must be written and strictly followed. The procedures for calibration must themselves be under statistical control, including equipment, procedures and operator skills.

A restricted area for storing and calibrating gages, test and measuring equipment should be located in an area where its integrity can be maintained. All gages should be checked out of the storage area and returned when no longer needed by Production. If a customer supplies special gages, they should be checked at intervals using the customers standards. If the customer has no standards, the gages should be checked according to appropriate NIST traceable standards.

Calibration procedures must be written and enforced. All gages and instruments need to be calibrated at least once each year. Obsolete or out-of-service gages and instruments must be tagged accordingly. Personal, as well as, company owned inspection tools should be properly and regularly calibrated. Recommended intervals for calibrating common equipment and gages is as follows:

1	- Gage blocks (working)	260 working days
2	- Gage blocks (masters)	260 working days
3	- Test indicators	260 working days
4	- Dial indicators except 0.0001"	260 working days
5	- Dial indicators 0.0001"	120 working days
6	- Micrometers	260 working days
7	- Calipers	90 working days
8	- Verniers	260 working days
9	- Smooth plug gages (Go/No Go)	60 working days
10	- Smooth ring gages (Go/No Go)	60 working days
11	- Smooth setting gages (all)	260 working days
12	- Thread gages, plug (working)	60 working days
13	- Thread gages, ring (working)	60 working days
14	- Thread gages (masters)	260 working days
15	- Electronic indicators	260 working days

For the purposes of this paper a "Working Day" is defined as any eight (8) hour shift during which the gages are used at any time.

CONTROLLING PROCESS BATHS

All baths must be analyzed on a regular basis. Scheduled analyses should be done at least weekly. If the need arises the bath could be analyzed on short notice. Wet chemistry analytical methods for bath control are recommended. This is simply because, while most modern instrumental methods are more accurate, they are also more

subject to error. Bath dilution for analysis is
usually done serially, each dilution compounds the
error. So the less precise wet method may well be
more accurate.

Initially, it's a good idea to analyze
the baths quite often, perhaps every hour to
determine how the bath varies with time and how
changes in the test results affect finished goods
quality. Suppliers are often very helpful in this
regard. After developing some history on a
particular bath its usually possible to analyze
less often. If it can be determined, as a function
of time, how much of a given chemical is being
consumed, it is then possible to make smaller adds
on a regular basis. This gives a more consistent
bath, and eventually allows analyzing less often.

SHIPPING IT RIGHT

The handling, packaging and shipping of
materials requires planning, control and
documentation as does any other operation. Marking
and labeling of materials must be legible, durable
and adequate to identify any particular batch in
the event of recall or should the need for special
inspection arise.

Non-conforming materials must not be
shipped unless the customer expressly approves
such shipment. When non-conforming material is
shipped the shipping documents should be clearly
marked to describe the non-conformance. When
non-conforming material is shipped it should bear
the name of the customer's agent who authorized
such shipment.

Every lot submitted for acceptance
inspection shall be kept as a unit, apart from
other lots and out of the normal flow of
materials. All production and inspection
operations must be kept in proper order, with each
step completed before the next step is begun.
Unidentified, or non-conforming, materials will be
removed from the normal flow of production until
it is properly identified, inspected and either
accepted or rejected. Reworked material must also
be segregated until its acceptance has been
verified.

Shipping is the last bastion of protection for the customer's interests that is under a company's control. Once shipping's job is done and the parts are gone there isn't much that can be done. It is therefore, critical, that no order be shipped until all shipping papers are in order, stamped and signed. This includes any required certifications, test reports, special samples, etc. The parts must also have been packed in accordance with customer requirements and checked for compliance. All materials should be handled with the utmost care at all times to prevent damage, deterioration and substitution. Proper customer identification on the packaging or parts prevents loss and misdirection of orders. In general, all items should be returned to the customer packaged in the same manner in which they are received, unless otherwise instructed in writing, by the customer.

DOING IT ALL ON SCHEDULE

Any company with a computer has the ability to control production schedules. One very simple way to accomplish this is by using a Data Base. While more elaborate forms of record keeping and scheduling are possible, this approach is both simple and convenient.

The following system is suggested:

Category: Customer
       Finish Code
       Date Received
       Date Due
       Shop Order Number
       Part Number
       Number of Parts
       Date Shipped
       Actual Turn Around

A Data Base is created using the Categories above. As parts are received they are entered into the Data Base. As a rule use a fixed "Turn Around" time in days for the "Due Date" unless the customer requests otherwise. Ten working days is a good "Turn Around" to begin with but any time may be used. The Data Base is then used as a simple scheduler. Each day the "Date

Shipped" Category is sorted for "is blank", thus
the only parts that will be printed out will be
those that are still in-house. Any parts that are
approaching the 10 day limit can be given the
highest priority. Properly used this is a very
simple and effective scheduling tool.

CONCLUSIONS

This paper is not intended to be
exhaustive, nor is it a checklist for compliance
with a specific set of requirements, it is rather
designed to help establish broad guidelines for an
effective Quality Management System (QMS). It is
intended to be a practical set of attainable,
doable elements that can help any metal finisher
achieve better quality and thus happier more
satisfied customers.

An effective QMS should be designed to
satisfy customer needs and expectations but must
also serve the company's interests. A well
structured QMS is a valuable management resource
for the optimization and control of quality in
relation to risk, cost and other benefit
considerations.

None of the above is particularly
difficult, nor does any of it require either
high-tech or higher math. Yet it is data driven.
Everything mentioned here is well within the
capabilities of any metal finishing shop. About
the only secret weapons needed are knowledgeable
production personnel who care.

THE GURU AND THE STANDARD - AN EMERGING CASE STUDY OF TQM IN
SPECIALTY LITHIUM BATTERY MANUFACTURING

William D. K. Clark

Wilson Greatbatch Ltd.
10000 Wehrle Drive
Clarence, New York 14031

ABSTRACT

Many companies have come to see the value of Total Quality Management
(TQM) as a driving force for their businesses. Each needs to customize the
approach to their company and its culture. The combination of a quality
improvement process based on the teachings of Philip Crosby and a
documentation structure using the ISO 9000 elements provides a means for
establishing a baseline for the quality system in the organization in
documenting the improvements as they are made. Examples illustrate how
the two systems interplay for a lithium battery manufacturer on a journey of
continuous improvement.

INTRODUCTION

Total Quality Management (TQM) as a driving force for business has been with us
for at least one to two decades. The companies that are true world competitors have been
aware of the need and value for at least this period. They have demonstrated its value in
remaining global companies that can compete in the world economy. Many companies in
the United States who have not ventured too far into world markets are seeing the value
from a defensive point of view as companies from overseas have entered their markets.

For companies whose business has a strong technology base, the principles of
TQM are particularly important as the discipline needed to make complex, technically
demanding products requires that all members of the company be effective contributors to
the development, manufacture and sale of the products. Each person must understand their
role in the company and have knowledge of tools and methodologies to improve the
operations and products of the company. For companies to have an effective TQM system,
they require a documentation structure and a management structure that support and foster
the efforts of the individuals in the company.

<u>History of the Company</u>

Wilson Greatbatch Ltd. (WGL) is the world's largest supplier of lithium batteries
for implantable medical applications. They also supply lithium batteries and precision
machined metal parts for other technically demanding environments. The company has
about 500 employees with operational capabilities ranging from fundamental R&D through
manufacturing and sales and marketing, all of which reside in one location.

The quality systems that the company has employed since the early 1970's had their basis in the Good Manufacturing Practices (GMP) of the U.S. Food & Drug Administration. (1) These quality practices focus entirely on the manufacturing aspect of the organization and say little or nothing with respect to the designing of the product or how the need for a product is identified or how to effectively get it to the customer and have it meet the requirements. The GMP as a quality system is ranked as an early system when compared to the systems that one has come to associate as state of the art TQM approaches (Figure 1).

The GMP system as employed by WGL has been able to meet one of the prime requisites of any TQM system, namely to deliver product that satisfies the customer requirements. However, in many cases this has been done at considerable internal costs. In an attempt to reduce these internal costs, a quality system based on a more global approach is in the process of being implemented. The selected initial approach uses the quality improvement methodology as espoused in the writings and teachings of Philip Crosby and a documentation structure for the processes in the company as outlined in the ISO 9000 quality standard. (2,3,4,5) This approach was selected as a manageable approach that fits with the culture of the company at this time. The approach also has the flexibility to allow the incorporation of other elements and ideas that exist or evolve in the developing arena of TQM and the spirit of continuous improvement.

The Crosby Elements

The elements of the Philip Crosby approach to quality improvement are documented in several books and embellished in course material available through Philip Crosby Associates, Inc.® The approach is structured on a basic view that all work is a process embodying customer-supplier relationships that one continually strives to improve. Crosby espouses 4 absolutes of quality dealing with the definition of quality, the system and the performance standard for the improvement of quality, and the measurement of quality. The 4 absolutes are shown in Table I.

Table I. The 4 Absolutes of Quality

1. The definition of quality is conformance to requirements.

2. The system for improving quality is prevention.

3. The performance standard is zero defects.

4. The measurement of quality is the price of nonconformance.

Crosby uses a quality improvement management structure that can overlay an existing functional organization structure to implement a 14 step process (Table II) that becomes the system of how the company manages total quality.

Table II. The Crosby Implementation Steps

1.	management commitment	8.	zero defects planning
2.	quality improvement team	9.	zero defects day
3.	education	10.	goal setting
4.	measurement	11.	error cause removal
5.	cost of quality	12.	recognition
6.	quality awareness	13.	quality councils
7.	corrective action	14.	do it all over again

The approach embodies many of the elements common to the approaches as espoused by the other well known quality gurus such as Dr. W. Edwards Deming and Dr. J. M. Juran. (6,7) This paper is not meant to discuss the comparison and relative merits of each of these approaches. Suffice it to say that each methodology has valuable points and the selection of a particular approach of one of the gurus is a good way to start an organization on the journey of TQM. As the organization gets further along the journey, it will customize whatever initial approach was chosen to reflect the business, technology and cultural needs of the organization.

The ISO 9000 Standard and Elements

The International Organization for Standardization has developed a series of quality system standards (known as the ISO-9000 series). The focus of the standards is to provide a list of basic elements that are needed to satisfy minimum requirements for a TQM system. (8) Adoption of the standards by participating countries and the companies resident therein creates the basis for a universal baseline of what a customer could expect from a supplier in terms of the quality of the product or service that is supplied. The adherence to the standards is monitored through an initial certification and lifelong semiannual auditing process by third parties. At present the standards have been adopted by over 90 countries worldwide. Such universal adoption will make these standards a requirement for doing business.

The ISO-9000 series contains five documents, of which three are the standards (9001, 9002, 9003) and the other two (9000, 9004) are an index and a set of guidelines. The 9001 standard is the most extensive as it covers companies that design and manufacture product. The 20 elements of the 9001 standard are listed in Table III.

An examination of the details of the elements shows a number of common features that focus on the considerations of responsibility, accountability and the issues that need to be addressed in that element. The wording in the standards does not provide a cookbook on how to address each element. The negative aspect of this feature is that work must be spent in developing a system to satisfy the requirement of the element, but the positive aspect is that it allows a flexibility in the approach taken.

Documentation forms the basis for satisfaction of the ISO elements and achievement of certification. However, the flexibility allowed by the standard permits one to follow the old quality documentation dictum of "say what you do and do what you say."

Table III. ISO 9000 Elements

1. management responsibility	11. inspection, measuring & test equipment
2. quality system	12. inspection & test status
3. contract review	13. control of nonconforming product
4. design control	14. corrective action
5. document control	15. handling, storage, packaging & delivery
6. purchasing	16. quality records
7. purchaser supplied product	17. internal quality audits
8. product identification & traceability	18. training
9. process control	19. servicing
10. inspection and testing	20. statistical techniques

If the documentation captures the former and you do the latter, then these two activities provide the basis for a high probability of a successful audit and subsequent certification. A recently published textbook offers pragmatic guidance on the implementation of the elements and on the role of the ISO elements as the basis for a TQM system. (8)

HOW THE GURU AND STANDARD WORK TOGETHER

The Documentation Structure

A tiered documentation structure is a preferred and natural approach (Figure 2). Documentation can take many forms that can include quality manuals, drawings, organization charts, forms, check lists, position descriptions, and equipment manuals. Much of this documentation exists in most companies, but the trick is to put it into a system that references from the top level down, minimizes changes to the documentation that need to be made if a change occurs in any given level, and provides a means to control how and when the changes are made. The documentation system provides the current record of how the processes in the company are to work and who has the responsibility for making them conform to the requirements.

The Quality Improvement Process

At the heart of the quality improvement process at WGL, is the cross-functional, cross-divisional group of ten individuals known as the Quality Improvement Team (QIT). While they take their strategic direction from the executive level management, they are the group that has the responsibility for the tactical management of the process. The QIT develops the system in which quality improvement occurs and removes roadblocks that the step implementation teams encounter. Each member has assigned responsibility to implement at least one of the 14 steps.

The ISO/QIP Synergy

Table IV depicts the areas of each system where there is commonalty of emphases. The ISO documentation system is the record of how things operate at present. This documentation by no means says or indicates that the operations are optimal. The QIP is the means to improve upon how things operate. In the true spirit of continuous improvement there will never be an optimum situation, but one which is always improving depending on what the requirements demand.

Table IV. QIP Implementation Steps Related to ISO 9001 Clauses

	ISO 9001 Clauses	Mgmt Commitment	QIT	Measurement	COQ	Quality Awareness	C/A	ZD Day Plan	Education	ZD Day	Goal Setting	ECR	Recognition	Quality Councils
4.1	Management Responsibility				D	D				D			D	D
4.2	Quality System	D	D	D	D	I	D	D	D	I	D	D		
4.3	Contract Review		D		I							D		
4.4	Design Control		D	D			D		I			I		
4.5	Document Control		D									D		
4.6	Purchasing		D	I	I		I					D		
4.7	Purchasher Supplied Product		I									I		
4.8	Product Identification and Traceability		D				D							
4.9	Process Control		D				I				I			
4.10	Inspection and Testing		I	D			I							
4.11	Inspection, Measuring and Test Equipment		I	D								D		
4.12	Inspection and Test Status		D											
4.13	Control of Nonconforming Product		D		D									
4.14	Corrective Action		D	I	I		D	I	D					
4.15	Handling, Storage, Packaging & Delivery		D									I		
4.16	Quality Records		D	D	D		D		D			I	I	
4.17	Internal Quality Audits			I		D	D	D		D		D		
4.18	Training		D		I	D	I		D				I	
4.19	Servicing		D		D									
4.20	Statistical Techniques		D	D	D	I	D				D			

Relationship: Direct = D, Indirect = I

SPECIFIC EXAMPLES FOR LITHIUM BATTERY MANUFACTURING

A limited number of examples will be presented to illustrate how a particular ISO element and the use of the QIP can combine to define and improve specific areas related to the development, manufacture and sale of specialty lithium batteries. This limited number of examples by no means conveys the scope of the areas susceptible to possible improvement as this approach applies to all processes in the company. Each example begins with the associated ISO element, and the QIP step's role is featured later in the text. The numbering of the ISO elements will be that for the U.S. equivalent known as the Q-90 series documents that can be obtained from the American Society for Quality Control.

<u>Ensuring the customer order gets filled correctly and on-time</u>

4.3 Contract Review. The supplier shall establish and maintain procedures for contract review and for coordination of these activities.

In this example, the company has a procedure that specifies how the order is to be entered and the information conveyed to the manufacturing floor. Opportunities for error arise when an order requires special terminations of the lithium cell not standard in a catalogue or the customer desires that the battery receive special qualification testing prior to shipment. Confusion or error in this process can be addressed by the use of the Error Cause Removal system whereby an employee can submit a request for action to this quality management system. The system is external to any functional grouping and requires a reponse by management. This system provides the means to address frustrations to which functional management has been slow to respond.

<u>Curing the over-the-wall syndrome</u>

4.4. Design Control. The supplier shall establish and maintain procedures to control and verify the design of the product in order to ensure that the specified requirements are met.

This is one of the major systems in a company that can be well documented and structured, but quite inadequate in meeting the need for new product in shorter time frames. The development of a new lithium cell is a complex process that requires the balancing of many factors such as the power capability and safety, energy density and manufacturability, as well as the conformance to unique shapes for custom designed devices. In the case where the existing system may involve a serial movement of the idea to reality from marketing to R&D to engineering to manufacturing with transition through parochial department barriers and all the delays and potential for error, one can use the corrective action system of the QIP to form a team to examine ways to shorten development times. Once the improved methodology has been developed, e.g. a more integrated approach involving interdepartmental teams, the process can be documented in the form of the ISO structure that specifies the responsibilities and relationships.

<u>I didn't know it was out of calibration!</u>

4.11 Inspection, Measuring, and Test Equipment. The supplier shall control, calibrate, and maintain inspection, measuring, and test equipment, whether owned by the supplier, on loan, or provided by the purchaser, to demonstrate the conformance of product to the specified requirements.

Companies generally have methodologies established that ensure that gages and electronic measuring equipment receive regular checks to ensure they are accurately taking the desired measurements. The suitability of the measurement and the proper utilization of the information can be assessed as part of the measurement activity. Deficiencies in ensuring that all new equipment enters the calibration as well procedures to address product that was found to be manufactured with out of calibration equipment can be dealt with as part of the ECR system. Capacities measured for the lithium cells that are in error can result in the product not delivering the expected run time for a device and thus create an unhappy customer, or in extreme conditions for implantable medical devices, can be a life threatening situation.

<u>Time wasters in R&D</u>

    <u>4.6 Purchasing</u>. The supplier shall ensure that purchased product conforms to specified requirements.

    R&D activity needs the rapid and reliable supply of often exotic materials. Non-standard metal alloys to serve as electrode current collectors in corrosive electrolytes are most often difficult to find. The purchasing function can be a valuable resource in the location and timely supply of these materials, freeing scientists and engineers to concentrate on the experiments to be run as opposed to chasing down suppliers. Documented procedures that define the roles of the purchasing agent and the scientist remove opportunities for confusion and wasted effort. Management commitment to doing it right the first time and systems to take corrective when that does not occur will remove the distraction from R&D for the procurement function.

<u>How to handle a hot cell</u>

    <u>4.15 Handling, Storage, Packaging, and Delivery</u>. The supplier shall establish, document, and maintain procedures for handling, storage, packaging, and delivery of product.

    Lithium cells are the highest energy density battery systems available, and when electrically or mechanically abused have the potential to rupture violently. As part of the improvement activities at WGL, driven by management commitment and a zero defects performance standard of no safety incidents with the product internal or external to the plant, measurement is done on the number of events that occur and corrective action in the forms of customer education are undertaken. Detailed procedures on how to handle cells to avoid electrical or mechanical abuse, and cleanup methods when events do occur, were developed and are available to customers.

CONCLUSIONS

The combination of a documentation structure and a quality management structure for quality improvement provide a means to establish a baseline for a company's quality efforts and to capture the progress that is made as it undertakes the journey of continuous improvement.

ACKNOWLEDGMENTS

I wish to thank C. Holmes, R. Rusin, and E. Takeuchi for their helpful comments on this manuscript.

REFERENCES

1. FDA 91-4179, Medical Device Good Manufacturing Practices Manual, 5th Edition, U.S. Department of Health and Human Services, Public Health Service, Food and Drug Administration.

2. P. B. Crosby, in Quality is Free, McGraw-Hill Book Company, New York (1979).

3. P. B. Crosby, in Quality Without Tears, McGraw-Hill Book Company, New York (1984).

4. P. B. Crosby, in The Eternally Successful Organization, New American Library, New York (1988).

5. ANSI/ASQC Q94-1987, American National Standard, Quality Management and Quality System Elements - Guidelines, American Society for Quality Control, Milwaukee, Wisconsin (1987).

6. W. E. Deming, in Out of the Crisis, Massachusetts Institute of Technology, Cambridge, Massachusetts (1982, 1986).

7. J. M. Juran, Juran's Quality Control Handbook, 4th Edition, McGraw-Hill Book Company, New York (1988).

8. S. C. Puri, ISO 9000 Certification - Total Quality Management, Standards-Quality Management Group, Washington, D.C. (1992).

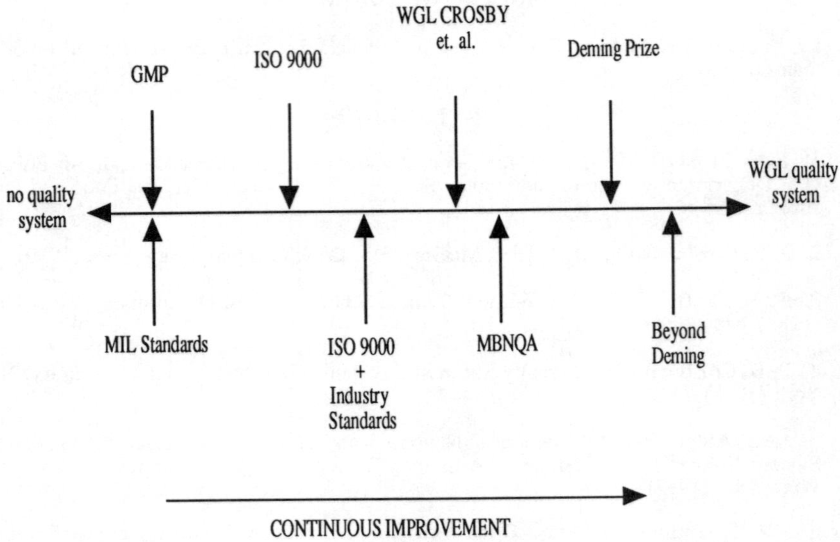

Figure 1. Hierarchy of TQM Systems

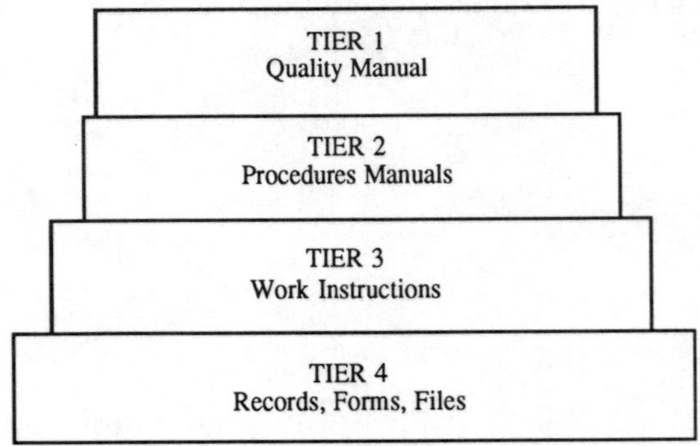

Figure 2. A Tiered Documentation Structure

ISO 9000 -- A CRITIQUE

Leslie W. Flott, C.Q.E., ASQC Fellow
Manager, Customer Satisfaction
Wayne Metal Protection Company
Fort Wayne, Indiana

Author, "Quality Control" column,
Metal Finishing Magazine

Abstract

ISO 9000, a Quality Management Systems standard whose popularity is sweeping the world, has executives asking if it can help against stiffening competition. The answer is "yes", but the system is flawed. ISO 9000 is a template designed to fit everyone's quality system needs, but can any system meet everyone's needs?

ISO 9000 isn't a complete quality system. It doesn't address questions of quality costs, process improvement or customer satisfaction. Companies shouldn't replace existing systems with the ISO 9000 criteria.

INTRODUCTION

In December 1992, one of the last phases of the process fashioned in the European Community White Paper, and the Single Europe Act amendment to the treaty of Rome (finalized at the Maastricht summit) was slated to become an actuality. The Maastricht Treaty was proposed to create one Europe, one currency, one market and one quality standard. As of this moment, the quality standard, called ISO 9000, has become an actuality while a united Europe is still very much in doubt.

What is ISO 9000? What in fact is (the) I.S.O.? The initials are simply an acronym for the International Organization for Standardization, a multi-national association of ninety-one mutually cooperating nations working toward a goal of world-wide quality assurance. Each subscriber nation has established a national third-party audit and registration scheme that is usually government sponsored. The American National

Standards Institute (ANSI) and the American
Society for Quality Control (ASQC) are the ISO
9000 sanctioning members for America.

ISO 9000 is very much in all of our
futures, and it is crucial that it be clearly
understood. According to Peter Burrows, in
Electronic Business dated January 27, 1992, "The
biggest challenge for U.S. firms has been the
confusion over ISO 9000. The guidelines are
complex, it goes by several names and there are
three levels of certification.: Among those of us
who have studied ISO 9000, some see it as a
blessing; others, like Mr. Burrows, as more of a
curse. It is neither. ISO 9000 is simply a series
of Quality Management Systems (QMS) standards
adopted in the Europe Community (EC) as EN-29000
in an attempt to assure uniform product quality
and reliability. In intent, it is not different
from MIL-Q-9858, Total Quality Management (TQM) or
the Deming approach. It is both more and less than
any of these, and shares features of all of them.

Yet in a way, this begs the question.
The explanation explains nothing. To more
completely answer the question of what ISO 9000
is, and what it means, we must take a small detour
into the past fifty, or so, years of European
political and economic history.

<u>MODERN EUROPEAN ECONOMIC HISTORY</u>

Just prior to World War II the sense of
"the world as community" was at an all time low in
Europe. Nationalism was rampant and nearly
everyone was nearly everyone else's enemy. The war
that inevitably followed destroyed Europe's
manufacturing base, which, coupled with a
discernible need to provide for the common
regional defense, gave birth to NATO and the
Warsaw Pact. French statesman Jean Monnet realized
that continent-wide cooperation was needed in
Europe, and the need for collective security would
drive the process of cooperation. This was the
first step toward the formation of what became the
formal European Community (EC). In 1950, following
the Berlin blockade, Monnet suggested to Minister
Robert Schuman, that France and West Germany pool
their coal and steel resources. In 1951, four
other countries joined them to form the European

Coal and Steel Community (ECSC), which became a
reality in 1952.

An important result of the ECSC was the
elimination of trade barriers between the member
nations, which they quickly seized upon as a
tremendous advantage. Still more important,
however, was the establishment of Europe's first
international governing institutions. By 1957, by
the treaty of Rome, Euroatom and the European
Economic Community (the EEC) also came into being.
The three organizations, Euroatom, ECSC and EEC,
shared the same legislature and judiciary bodies
but had separate executive agencies. The one great
weakness was that it required an unanimous vote of
the member nations to pass any legislation.

Thus the Treaty of Rome brought the six
nations of ECSC -- Belgium, France, Italy,
Luxembourg, the Netherlands and West Germany --
into a cooperative trade association for the
purpose of forming a single continent- wide
market. This "common market" was to be free of
barriers to movement of goods, people, services,
and money. The EEC came into official existence in
1958, with the ECSC-member nations as charter
members.

By 1967, the three separate
organizations -- Euroatom, ECSC and EEC --
formally mingled their executive branches, thus
for the first time creating a kind of United
States of Europe, the European Community (EC).
Jean Monnet's dream had become a reality. Note
that the European Economic Community (the EEC) and
the European Community (the EC) are not exactly
the same entities; however, common usage often
blurs the differences. The terms are used
interchangeably.

As other nations became convinced of the
potential value to them of the EC, the EC expanded
further. First to join the charter members were
Denmark, Ireland and the United Kingdom in 1972.
The addition of these three nations to the EC made
the EC larger than the United States. Greece,
following years of revolution and internal unrest
was next to join in 1981. The unexpected collapse
of Communism in the 90's and the dissolution of
the Warsaw Pact has opened the EC to still further
expansion.

One of the most singular events of this
period was the issuance of the "EC White Paper" in
1985. This history-making document stated that for
the EC to become a single free market, its rules
had to be enforceable. The "EC White paper" is
essentially a plan for achieving this goal. By
December of 1992, one hundred-eighty of the
initial 279 directives included in the "EC White
Paper" had been adopted.

In 1987, the first amendment to the
Treaty of Rome, the "Single Europe Act" was
adopted. One facet of this act was to formally
define Europe as a single market without internal
frontiers. But perhaps the most important
provision of the Act was that it abrogated the
need for an unanimous vote of the members
substituting a rule that requires only a simple
majority for purposes of passing legislation. This
single step has greatly accelerated the process of
change in Europe.

Where does the EC go from here? Will we
see the historic recreation of "Fortress Europe"
along economic lines? Will the EC come to dominate
world trade? Will they play the part of economic
dictators, or will they promote a more democratic,
laissez faire, egalitarian world economy? There is
no doubt that the EC will become a major player in
the years to come. Only time will tell what part
they will play. Regardless of other factors the EC
represents one of the world's biggest growth
markets.

The EC is a major economic unit with
more people, and produces more goods than the
United States. They also import more than any
country on earth, and their major trading partner
if the United States. The Ec has increased the
real wealth of its citizens continuously since the
1950's. Concurrently with growth was the growing
perception of the need to establish the quality of
EC produced goods as second to none in the world.
By 1970, an international quality systems standard
was being discussed in the EC.

THE EVOLUTION OF ISO 9000

Some years before their European
cousins, the Americans had recognized the need to

establish a "National Quality Management Program".
This aspiration resulted, in 1959, in MIL-Q-9858,
titled "Quality Program Requirements." This was
part of an increasing demand for more effective
quality management practices that was, and is,
world wide. MIL-Q-9858 was revised in 1963 and was
adopted by Nato in 1968 as Allied Quality
Assurance Publication 1 (AQAP-1). Then in 1970,
the British Defense Ministry adopted the
provisions of AQAP-1 as their Management Program
Defense Standard, called DEF/STAN 05-8.

DEF/STAN 05-8 was further revised in
1972 to conform to the provisions of the then
proposed ISO 9000 and took on the new designation
DEF/STAN 05-21, -22, -23, and -24, which
provisions are themselves closely related to the
basic standards that would eventually make up ISO
9000, namely ISO 9001, 9002, 9003 and 9004. In
1979, the British Standards Institute used these
predecessor standards to develop the world's first
successful Quality Management System (QMS)
standard, designated BS-5750. Eight years later in
1987, the International Standards Organization
(ISO) published ISO 9000 containing most of the
provisions of BS-5750. Later that same year, the
two standards were reconciled, making them
essentially identical document. In that same year,
the EC adopted this standard as EN-29000, and the
ANSI/ASQC jointly published Q-90, also virtually
identical to ISO 9000. So in that banner year, all
the major QMS standards world-wide became
effectively the same document.

Even Japan has come up with it's own
version of ISO 9000. Officially adopted in October
of 1991, Japanese Industry Standard (JIS) Z 9900
gives the standard an almost global acceptance.

ISO 9000 CERTIFICATION AND REGISTRATION

American companies striving to remain
competitive in the global marketplace have already
begun to put into place quality management systems
that are intended to insure successful
certification under the guidelines of Q-90, or ISO
9000. The standard is intended to focus on the
recognized universal demand for quality assurance.
The strategy is to establish a mechanism whereby a
purchaser in one country can be assured of the

quality system of a supplier without having to do
an audit.

As with any new venture, successful
certification begins with the support, commitment
and involvement of top management. This is
important since management must understand exactly
what the commitment means in terms of dollars,
before the process can begin. Half-hearted
attempts at ISO 9000 certification and
registration are costly and are doomed to failure.

The various component standards that
make up ISO 9000 are essentially a set of
Guides-to-Practice for target organizations. ISO
9000 is simply a guide to the proper selection of
one of the other standards. ISO 9001 establishes
criteria for Quality Assurance for designers,
manufacturers and providers of products and
services. ISO 9002 establishes criteria for those
not actually involved in the design, but is
otherwise similar to 9001. ISO 9003 is a model for
final inspection and test practices, primarily but
not limited to those who are distributors and
added value providers. ISO 9003 has important if
somewhat limited application. Finally, ISO 9004 is
an operational guidebook for application of the
management and quality systems elements of a QMS.

THE CERTIFICATION METHODOLOGY

Certification & Registration under ISO
9000 involves a number of essential steps. These
represent the minimum steps necessary to achieve
certification and registration. The steps are:

```
 1 - Get complete top management agreement
 2 - Define the Quality Manual
 3 - Implement the policies in the Manual
 4 - Train all personnel in needed skills
 5 - Define all necessary procedures
 6 - Perform a thorough internal audit
 7 - Employ standard operating procedures
 8 - Hold a 3rd party preassessment audit
 9 - Make corrective actions/improvements
10 - Submit the Manual to the Registrar
11 - Conduct an assessment visit
12 - Take corrective actions and reassess
13 - Registration
14 - Celebrate
```

15 - Schedule semi-annual audits

BENEFITS OF CERTIFICATION AND REGISTRATION

Significant benefits accrue to those who successfully achieve ISO 9000 certification and registration. Such companies attain world class status, with world-wide stature. Prospective customers, both domestic and foreign, simply need to confirm ISO 9000 certification and registration to assure themselves of the supplier's consistency, control and assurance of quality.

Unfortunately while the EC's intent was to establish the ISO 9000 series of standards as a sort of minimum universal quality system, what has happened is that each subscriber nation has adopted its own certification and registration scheme. In most countries this is done through a governmental agency, this is not the case in the USA. The assorted schemes only add to the complexity of the whole process.

Beyond the EC things get even more complicated. At present the ISO 9000 initiative is market driven, but it may some day have the power of law. Regulatory uniformity and the necessary controls are already being discussed.

While certification and registration can and will help a company's marketing efforts, the greatest benefit may well be the audit process itself. Companies cannot help but benefit from comparing their internal quality systems against recognized world standards. The question may be; "Will all national certification and registration schemes be legitimate?"

THE ROLE OF ISO 9000

ISO 9000 is clearly not without critics, or possible short comings. Given the nature and history of the development of ISO 9000, its rapid evolution and attempts of apologists to apply it universally, not to mention that some people simply don't understand it, for it not to have pitfalls would be miraculous.

The very concept of ISO 9000 should cause us to approach it with some degree of caution. It

was created to be a universal template upon which
to design everyone's quality management system. It
is precisely because of its universality, that we
should approach it with caution. It would be folly
to assume that any program could meet every
possible quality need for every possible
organization in the world.

There is real danger that organizations
may abandon systems that are already in place and
working simply because those in charge think that
because ISO 9000 is the latest and greatest, it
must be the only way to go. As Subash C. Puri,
Chief Statistician of Agriculture Canada (see
"Deming + ISO 9000 -- A Deadly Combination for
Quality Revolution," 1991 -- ASQC Quality Congress
Transactions -- Milwaukee) points out clearly,
"The focus of attention of the manufacturing world
has been side-tracked. This is clearly creating
all kinds of apprehensions, ranging from confusion
to jubilation."

Please note that unfortunately, there
are organizations today that claim to offer ISO
9000 certification and registration but who have
not been officially sanctioned to do so. Many of
these are not in compliance with the well-defined
parameters of the EC initiative, or the
International Organization for Standardization.
There is even some question about the validity of
the American National Standards Institute (ANSI)
and the American Society for Quality Control
(ASQC) as the ISO 9000 sanctioning members for the
United States of America. The US is the only major
country that does not sanction through an agency
of the government, hence the question.

Companies attempting to secure ISO 9000
certification are cautioned to chose their
certifying company carefully. make sure that
whoever offers the certification are themselves
accredited through one of the existing
organizations. Because of the doubt raised above,
many American companies are actually seeking dual
certification.

DEMING Vs.ISO 9000

Most people who have even a passing
acquaintanceship with quality control concepts are

aware of W. Edwards Deming, the man who turned
Japan around in the 50's and 60's. Japan's top
quality prize is called the Deming Prize. Deming
is perceived by many as more philosopher than
engineer. His Fourteen Points, first published in
1984, are more thought provoking guides than
concrete hand-on action items.

Deming's Tenth Point says to eliminate
exhortations, slogans and targets. ISO 9000 is
clearly an identifiable target. I don't believe
that Dr. Deming is referring to this type of goal;
after all, a system based on his own philosophy
could itself be considered a target. The Eleventh
Point says to eliminate numerical goals and quotas
for the workforce and management. Again, Deming
seems to run contrary to ISO 9000, which does set
rigid goals.

I am neither endorsing nor refuting Dr.
Deming or his philosophy, merely presenting his
ideas for the purpose of comparing some of his
thoughts to ISO 9000. Those interested in
obtaining more information about Deming's Fourteen
Points should contact ASQC Quality Press,
Milwaukee, Wisconsin, for any of dozens of books
on the subject. Dr. Deming has, I think,
unfortunately, often severely criticized both the
Malcolm Baldrige National Quality Award criteria
and ISO 9000. At least a part of Deming's critical
view stems from his Fourteen Points.

One great advantage of Deming's approach
is that, unlike ISO 9000, it is flexible. Some
apologists for Deming, and various other systems
of quality improvement in current vogue, seem at
times to attack ISO 9000 a bit too vigorously,
comparing it unfavorably to their own "pet"
theory. ISO 9000 is not in competition with other
quality improvement systems, but should augment
them.

PHILOSOPHICAL DEBATES

ISO 9000 and its related quality
assurance standards were created to help persuade
a customer, or potential customer -- or a third
party -- that an organization is working to the
highest quality standards. Does ISO certification
and registration guarantee that this is true? The

mere fact that such a question can be asked should
make people wary of the effectiveness of trying to
achieve quality through systems like ISO 9000
alone. Simply complying with ISO 9000's terms
isn't enough. The total quality philosophy implied
by ISO 9000 must be put into practice.

Often such systems simply
institutionalize a bureaucratic and ineffective
approach that not only fails to encourage, but may
well stifle, improvement. The objective of an
organization's quality program becomes one of
satisfying the standard rather that perfecting the
product or service. I repeat, the total quality
philosophy implicit in ISO 9000 must be put into
practice. Anything less is corporate and perhaps
national suicide.

Critics of ISO 9000 are numerous, among
the loudest is, of course, Dr. Deming. In addition
he also criticizes the Malcolm Baldrige Award
criteria, Phil Crosby and Armund Feiganbaum, and
nearly everyone else other than Deming. The
problem is that many people have a hard time
getting beyond the Deming philosophy, as outlined
in his Fourteen Points. Many pragmatists out there
are looking for a more concrete do-it-yourself
manual. Many people have a hard time dealing
exclusively with the philosophy.

Puri and others who do, apparently
understand Deming, compare ISO 9000 unfavorably to
the Deming philosophy. His system is rather
scholarly for some, but it is one of several very
workable systems by which high quality can be
achieved. What we are dealing with is a difference
in philosophies. Deming's is simply not the only
approach to ultimate quality and continuous
process improvement.

There are many ways of achieving better
quality. Total Quality Management (TQM), ISO 9000,
the Deming Philosophy, Crosby's Fourteen Points
all of them work. To assure improvement
organizations must gather, evaluate and act on
facts not assumptions. Pareto analysis,
Statistical Process Control (SPC), Ishkawa's Cause
and Effect Diagrams, Quality Function Deployment,
runs charts, Failure Mode and Effect Analysis

(FEMA) are all tools for improving process quality.

ISO 9000 certification and registration are valuable tools and like any good tools must be used for the purpose for which they were designed to be most effective. Their intent is to allow potential customers to prequalify suppliers by confirming that the supplier meets certain quality management system norms. In this context, they have great merit and are worthwhile. However, certification and registration in no way guarantees satisfactory products and services.

ISO 9000 only verifies that a company knows how to produce good products, assurance of such performance requires looking deeper. ISO 9000 must not be looked upon as the end of a quality process, but as a fine beginning.

Much of the material used in this paper originally appeared in the March and April 1992 issues of <u>Metal Finishing</u> magazine. Special thanks to Elsevier Publishing Co., publishers of <u>Metal Finishing</u> magazine, and Mr. Mike Murphy, Editor, for allowing the author to use the material in this paper.

DEVELOPMENT OF
HIGH TEMPERATURE AND HIGH STRENGTH
HOT DIP ALUMINIZED STEEL SHEET

Yasutaka Kawaguchi, Kyosuke Ohasi
Susumu Yasuhara
Coated Steel Department Sakai Works
Nissin Steel Co., Ltd.
NO.5 Ishizu Nishi-Machi Sakaishi, Osaka,
592 Japan

ABSTRACT

Hot-dip aluminized steel sheets used in automotive exhaust system parts must be strong enough to withstand vibration and external force at high temperatures. In response to this demand, a new material was developed by hot-dip aluminizing steel sheets that contain 0.6% silicon (Si) and 1.0% manganese (Mn). However, the poor wettability of this steel sheet's hot-dip aluminum made it easy for defects to appear in the coating, and adhesion between the coating layer and base steel was also poor.

This time, these Si-Mn steel sheets were hot-dip aluminized after first electroplating them with at least 1.5 g/m^2 of an iron-boron (Fe-B) alloy with a minimum 0.001% boron(B) composition ratio thus succeeding in the development of a high-temperature and high-strength hot-dip aluminized steel sheet with superior corrosion resistance and heat resistance.

INTRODUCTION

Aluminized steel sheets that have hot dipped in an aluminum alloy with an 8 to 12% silicon content have superior corrosion resistance, heat resistance, heat reflectivity and light reflectivity. And they are widely used in such applications as automotive exhaust system parts, heat reflector plates in kerosene stoves and other combustion devices, light reflector plates in lights, and other household electrical appliances.

Recent years have seen a growing demand for automotive exhaust system parts that are not only corrosion-resistant but also strong enough to withstand vibration and external force at high temperatures. After discovering that adding silicon (Si) and manganese (Mn) to steel is effective in achieving these qualities, steel sheets containing 0.6% silicon and 1.0% manganese were developed (1). The composition of these steel sheets reflects target strength levels (i.e., a tensile strength of 392 to 441 N/mm2 at room temperature and 147 to 196 N/mm2 at 600° centigrade). However, directly hot-dip aluminizing these steel sheets resulted in poor wettability in the hot-dip aluminum, leading to bare spots (i.e., flaws in plating) and poor adhesion between the coating layer and base steel.

In view of this, and with the goal of improving the adhesion of the hot-dip aluminizing without losing the superior qualities of the Si-Mn steel sheets, a high-

temperature and high-strength hot-dip aluminized steel sheet was developed by electroplating the lower of the hot-dip aluminum coating with an Fe-B alloy.
The following is a description of the manufacturing conditions of these steel sheets as well as how the adhesion of the hot-dip aluminum coating was improved.

EXPERIMENTAL PROCEDURE

Specimens

1.5 mm thick cold-rolled Si-Mn steel sheets were used as the specimens, the composition of which is shown in Table I.

Conditions for Fe-B alloy electroplating

The aforementioned steel sheets were electroplated with the Fe-B alloy with the plating solution composition and conditions shown in Table II. The plating weight of the Fe-B alloy electroplating was adjusted by changing the electroplating time. And the B composition ratio of the Fe-B alloy electroplating was adjusted by changing the amount of boric acid in the electroplating solution.

Conditions for hot-dip aluminizing

Figure 1 shows the conditions for hot-dip aluminizing. The Fe-B alloy-electroplated steel sheets were preheated for 30 seconds at 800° centigrade in an atmosphere of 50 vol.% H_2-N_2 atmosphere gas, after which they were immersed for two seconds in a solution of aluminum and silicon (9.5%), that was heated in the same atmosphere to 670° centigrade, thus fabricating hot-dip aluminized steel sheets with a coating weight of 50 g/m^2 per side.

Assessing the adhesion of the Fe-B alloy electroplating

After completely contact-bending these Fe-B alloy-electroplated steel sheets, tape was applied to the bend and then removed. Peeling in the Fe-B alloy electroplating that adhered to the tape was evaluated and rated according to the following three classifications.

○ : No peeling
△ : Partial peeling
✕ : Complete peeling

Assessing the adhesion of the hot-dip aluminum plating

The amount of specked coating defects on the surface (a 50 x 100 mm area) of a steel sheet electroplated with an Fe-B alloy and then hot-dip aluminized was assessed according to the following five classifications.

Level 5: No specked coating defects
Level 4: No more than five specked coating defects less than 1 mm in diameter
Level 3: More than five specked coating defects less than 1 mm in diameter
Level 2: Many specked coating defects less than 1 mm in diameter
Level 1: Many specked coating defects more than 1 mm in diameter

RESULTS AND DISCUSSION

Conditions for optimizing Fe-B alloy electroplating

Figure 2 shows the effects that Fe3+ concentration and pH in the plating solution had on the adhesion of the Fe-B alloy electroplating. As it shows, adhesion between the Fe-B alloy electroplating layer and the Si-Mn steel improves when electroplating is performed with a maximum Fe3+ concentration of 10 g/l and a maximum pH of 1.50.

It is possible that the reason electroplating adhesion becomes poor with high Fe3+ concentrations is that iron hydroxide is drawn into the electroplating layer.

The reason that electroplating adhesion decreases when Fe3+ concentration is high is shown in Figure 3, which shows the effects of plating solution pH on the adhesion and current efficiency of Fe-B alloy electroplating. Generally, during electroplating, hydrogen gas forms on the steel sheet as it is plated with metal ions. This hydrogen gas serves to remove, through reduction, the oxide film that forms on the steel sheets being electroplated. However, because raising the pH lowers the concentration of hydrogen ions in the electroplating solution, less hydrogen gas forms on the surface of the steel sheets, and, as a result, less reduction occurs and adhesion of the plating decreases.

The results, above, show that by controlling the concentration of Fe3+ in the plating solution to within 10 g/l and pH to within 1.50, Fe-B alloy electroplating with superior adhesion is possible.

An investigation of the optimum plating weight of the Fe-B alloy electroplating and the optimum B composition ratio of the plating layer

Figure 4 shows the effects of Fe-B alloy electroplating weight and B composition ratio of the plating layer on the adhesion between Si-Mn steel and hot-dip aluminum. The results in Figure 4 shows that steel sheets with superior adhesion with the hot-dip aluminum can be fabricated by electroplating Si-Mn steel with a minimum of 1.5 g/m^2 of Fe-B alloy containing at least 0.001% boron by weight.

Photo 1 shows the difference in surfaces, after hot-dip aluminizing, produced by Fe-B alloy electroplating. In contrast to the bare spots here and there on the Si-Mn steel sheet not electroplated with the Fe-B alloy, none are seen on the Si-Mn steel sheet electroplated with 1.5 g/m^2 of Fe-B alloy.

The reasons for this are explained using the Si-Mn steel in Figure 5 and a reaction model of hot-dip aluminizing. The Si-Mn steel that is not electroplated with the Fe-B alloy, silicon diffuses and thickens on the surface layer of the steel sheet during short-term annealing in the atmospheric gas (H$_2$.N$_2$) prior to hot-dip aluminizing, thus causing an oxide film to form [2][3]. This oxide film, which is comprised mainly of SiO$_2$, lowers the adhesion of the hot-dip aluminum. The reason that electroplating the Si-Mn steel with the Fe-B alloy improves adhesion of the hot-dip aluminum is that the Fe-B alloy electroplating layer acts as a barrier to silicon that would otherwise diffuse and thicken on the surface layer of the steel sheet. When the Fe-B alloy electroplating layer is too thin, however, the silicon passes through the barrier of the Fe-B alloy electroplating layer, then diffuses and thickens, thus lowering the adhesion of the hot-dip aluminum. As shown in Figure 4, this is why an Fe-B alloy electroplating layer of at least 1.5 g/m^2 is necessary.

In addition, the reason that coating adhesion of the hot-dip aluminum improves when the B composition ratio of the Fe-B alloy electroplating layer increases is, it is felt, because the uniform electrodeposition of the Fe-B alloy electroplating is improved.

APPLICATION WITH NO.1 GAL

Figure 6 shows an outline of No.1 GAL (Continuous Galvanizing and Aluminizing Line), a production line that produces 34,000 tons of hot-dip aluminized steel sheet and hot-dip galvanized steel sheet per month.

Beginning with the entry side, this line is comprised of the cleaning section, where oil on the surface of the steel sheets is removed; the annealing furnace, where the steel sheets are annealed; the aluminum or zinc pot, where hot-dip coating of the steel sheets is performed; the skin pass mill, where the shape of coated steel sheets is modified; and the chromating section, which imparts the steel sheets with corrosion resistance.

Based on the experimental results, above, Fe-B alloy electroplating equipment was installed at the entry side of No.1 GAL's annealing furnace and used in the fabrication of high-temperature and high-strength hot-dip aluminized steel sheets.

Figure 7 shows an outline of the Fe-B alloy electroplating equipment, which, beginning from the entry side, is comprised of the sulfuric acid pickling section, rinse section, the Fe-B alloy electroplating section, rinse section, and dryer. The composition of the Fe-B alloy electroplating bath was controlled in accordance with data obtained in experiments, while the system was also designed so that the plating weight of the Fe-B alloy electroplating sufficiently satisfies the requirement of 1.5 g/m^2 per side.

As shown in Figure 8, this equipment application resulted in primary yields for high-temperataure and high-strength hot-dip aluminized steel sheets far higher than yields obtained with steel sheets not electroplated with this Fe-B alloy.

CONCLUSION

A high-temperature and high-strength hot-dip aluminized steel sheet with a base steel containing silicon (0.6%) and manganese (1.0%) was developed for use in automotive exhaust systems. To counter the effects of the diffusion and thickening of silicon that would otherwise lower the adhesion of the hot-dip aluminum plating, the lower layer of the hot-dip aluminum plating player is electroplated with an Fe-B alloy.

Using results obtained in experiments, Fe-B alloy electroplating equipment was installed on No.1 GAL:, thereby achieving yields far higher than possible with conventional steel sheets not electroplated with this Fe-B alloy.

REFERENCES

Journals

1. T. Yamada, N.Sakai and H.Kawase, TETU-TO-HAGANÉ., S477 Vol. 71 (1985)
2. T. Fukuzuka, N.Urai and K.Wakayama, TETU-TO-HAGANÉ, S494 Vol. 66 (1980)
3. U. Hirose, H.Togawa, and J.Sumiya, TETU-TO-HAGANÉ., p.665 Vol. 68 (1982)

Table I Composition of specimens (Wt.%)

C	Si	Mn	S	Ti	Al	Ni	Cr
0.01	0.632	1.01	0.016	0.267	0.04	0.02	0.03

Table II Condition of Fe-B alloy electroplating

Chemical composition in the plating bath	Total Fe ($Fe^{2+}+Fe^{3+}$)	50g/l
	Fe^{3+}	0 ~ 10g/l
	Na_2SO_4	80g/l
	Tartaric acid	1g/l
	Boric acid	5 ~ 50g/l
	pH	1 ~ 1.7
Condition of plating	Bath temperature	50 °C
	Current density	50 A/dm²
Fe-B plating weight		0~1.5 g/m²
B composition ratio of plating layer		0.0001~1.0 %

Anealing for 30 seconds at 800°C
in 50% N_2 - 50% H_2 atmosphere gas

Immersing for 2 seconds in Al bath
(including 9.5% Si) at 670°C

Fig. 1 Condition of hot dip aluminum

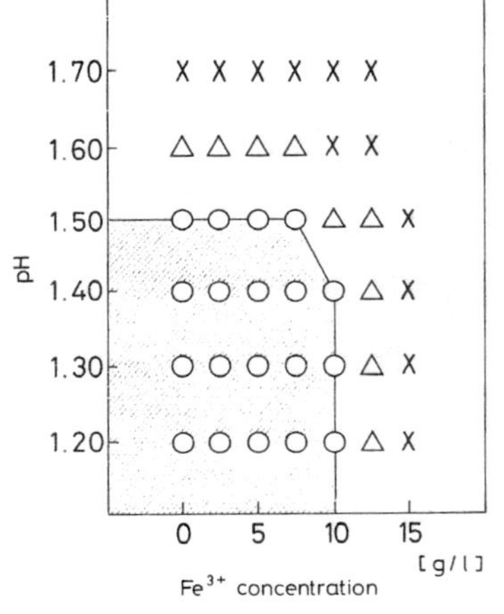

Fig. 2 Effects of Fe^{3+} and pH
in plating bath on
adhesion of Fe-B Plating

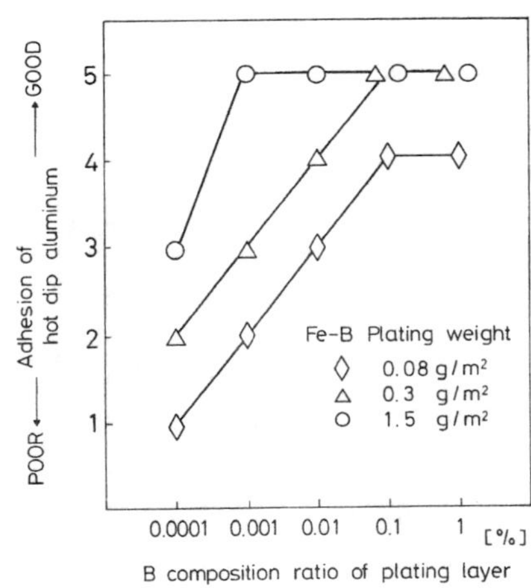

Fig. 4 Effects of Fe-B plating weight
and B composition ratio of
plating layer on adhesion of
hot dip aluminum

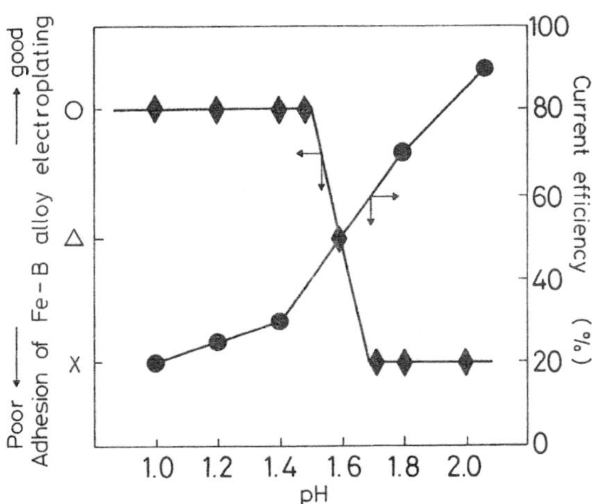

Fig. 3 Effects of pH on adhesion of
Fe-B alloy electroplating
and current efficiency

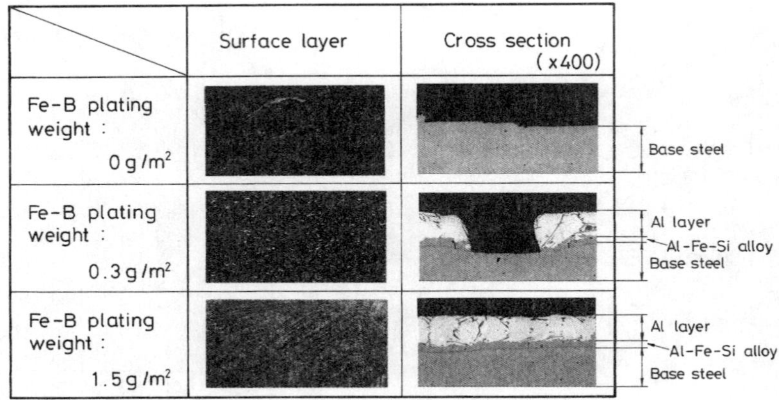

Photo. 1   Surface layer and cross section of Si-Mn steel after Fe-B alloy electroplating and Al-9.5%Si coating

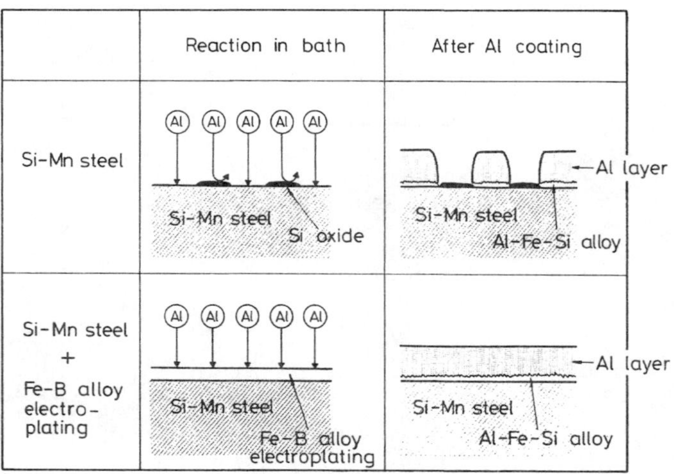

Fig. 5   Reaction of Si-Mn steel in Al bath

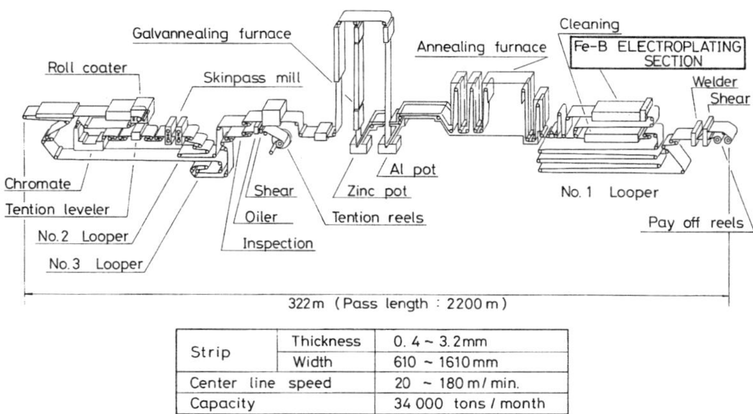

Strip	Thickness	0. 4 ~ 3.2 mm
	Width	610 ~ 1610 mm
Center line speed		20 ~ 180 m / min.
Capacity		34 000 tons / month

Fig. 6   Out line of No. 1 GAL

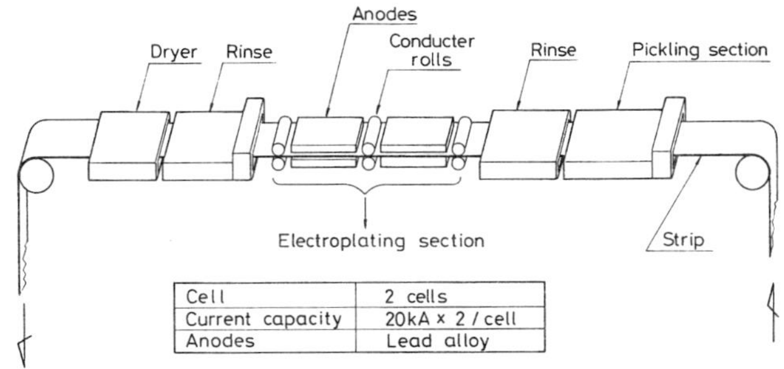

Cell	2 cells
Current capacity	20kA × 2 / cell
Anodes	Lead alloy

Fig. 7   Out line of Fe-B alloy electroplating section

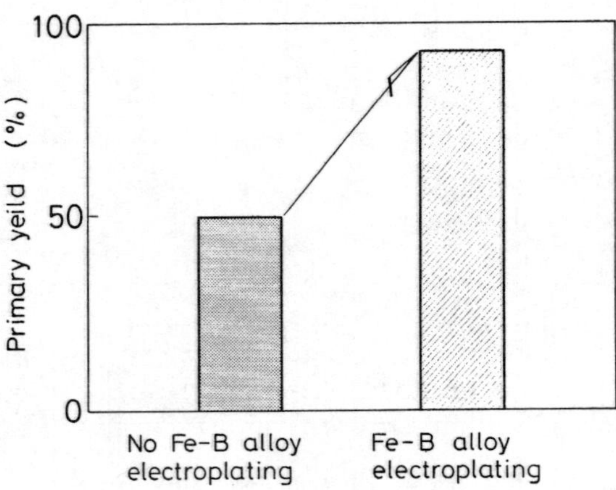

Fig. 8　Primary yeild of high temperature
and high strength hot dip
aluminized steel sheet

114

ULTRA-CLEAN TECHNOLOGY DEVELOPMENT OF STAINLESS STEEL PIPE FOR GAS DELIVERY SYSTEM IN SEMICONDUCTOR MANUFACTURING

Toshiki Kadonaga, Hirofumi Hamada and Hiroshi Ueda
Tubes & Pipe Technology Department, Kobe Steel, Ltd.
8-2, Marunouchi 1-Chome, Chiyoda-ku, Tokyo 100, Japan

Hiroshi Sato and Haruo Tomari
Surface Design & Corrosion Research Section,
Material Research Laboratory, Kobe Steel, Ltd.
5-5, Takatsukadai 1-Chome, Nishi-ku, Kobe,
Hyogo 651-22, Japan
and
Shigeharu Nakamura and Yoshihisa Kiryu
Technical Section, Chofu-Kita Plant, Kobe Steel, Ltd.
13-1 Chofu Minato-machi, Shimonoseki, Yamaguchi 752,
Japan

Electro-polished stainless steel pipings are widely used for gas delivery systems in semiconductor manufacturing. Because of increasing semiconductor device integration, quality requirements for these components become higher and higher. Characteristics of the required quality are particle free, out-gas free, and corrosion free. Recently, in highly advanced device manufacturing, gas impurity must be controlled below the ppb level, so it was found that even very small quantities of released moisture from pipe surfaces deteriorate the gas quality, and greatly affect device production yields. Thus, oxygen passivated pipe and Cr-enriched oxygen passivated pipe were developed.

As a part of TQC-Activity, these developments have been performed. This paper shows pipe manufacturing technology, features of the products, and TQC-activity.

BRIGHT-ANNEALED PIPE (BA PIPE) AND ELECTRO-POLISHED PIPE (EP PIPE)

In the semiconductor manufacturing process many kinds of gases, such as inert gas, corrosive gas, and poisonous gas are used for dilution, cleaning, and etching. With the increasing of device integration such as ULSI, super-fine pattern etching has become necessary, and the quality requirements for these gases have become higher. As these gases are distributed to the point-of-use through pipe lines, demands for surface quality such as roughness, adhesion, and residual particles, as well as weldability, workability, out gas characteristics and corrosion resistance quality have become severe for gas delivery pipes in semiconductor manufacturing.

Until now, bright-annealed stainless steel pipe (BA pipe) had been used for gas delivery systems in semiconductor manufacturing. However, BA pipe did not satisfy customer's quality requirements from the point of view of surface roughness, adhesion, and residual particles. Therefore elctro-polished pipe (EP pipe) was developed. EP pipe is electro-polished in the inside surface, its surface roughness is $0.2 \sim 0.3$ μm Rmax, and it is cleaned with hot pure-water in clean-room class 1000.[1], [2], [3] On the other hand, surface roughness of BA pipe is $1 \sim 2$ μm. EP pipings are widely used in $1 \sim 4$ M DRAM semiconductor manufacturing gas delivery systems, and BA pipings have been restricted to instrumentation dry air or exhaust gas pipe line.

Fig. 1 shows the manufacturing processes of BA pipe and EP pipe. Fig. 2 shows results of the particle test and Photo 1 shows SEM micrographs of inner surface of each.

OXYGEN PASSIVATED PIPE (OP PIPE)

Recently, in highly advanced device manufacturing plants, gas impurity must be controlled below the ppb level, and it was found that even very small quantities of released moisture (water) from the stainless steel pipe surface deteriorate the gas quality, and greatly affect device production yields. Thus, oxygen passivated pipe (OP pipe) was developed. OP pipe is an EP pipe which has been oxygen-passivated with ultra pure and dry oxygen and Ar gas in the inside surface of EP pipe. The surface film consists of oxide with enriched Fe and Cr, to a thickness of $100 \sim 200$Å ($10 \sim 20$ nm), and its surface looks golden colored. The volume of released moisture from the OP pipe surface is very low compared with that of EP pipe.[1], [4]

Fig. 3 and 4 show depth-composition profiles in the surface films of EP pipe and OP pipe. The thickness of oxide film of EP pipe is about 15Å, while that of OP pipe is 150Å. Fig. 5 shows the time dependence of moisture concentration in N_2 gas passed through pipe at room temperature. The volume of released moisture from OP pipe surface is 1/10 to that of EP pipe.

CR-ENRICHED OXYGEN PASSIVATED PIPE (CRP PIPE)

Furthermore, we found that the oxide film composition of the inside surface became only Cr oxide depending on the oxygen passivation treatment condition. For example, its treatment condition is 1 ppm ~ 1% H_2O or O_2 gas + 10% H_2 and ballanced with Ar gas, 500°C, 1 Hr. The Cr-enriched oxygen passivated pipe (CRP pipe) has the feature of easy moisture release, the same as OP pipe, and good corrosion resistance even to HCl or HBr gas containing water. Ni-based special alloys such as Hastelloy-C have catalytic activity and easily decompose SiH_4 gas in low temperature, but CRP pipe is more stable than Hastelloy-C to SiH_4/Ar gas.

Fig. 6 shows the depth-composition profile in the surface film of CRP pipe. The inner surface has only Cr oxide, and its depth is about 200Å. Photo 2, Fig. 7 and 8 show the results of corrosion test to HCl, HBr gas containing water, and Fig. 9 and 10 show the results of thermal decomposition test of SiH_4 and B_2H_6 gas. [5]

TQC-ACTIVITY

Tubes and pipe department of Kobe Steel had introduced TQC in 1982, and been awarded the 1989 Deming Prize. This paper shows activities of new products development as a part of TQC at that time and after that. Kobe Steel introduced TQC to improve the company quality for lasting prosperity and to supply the products and technology which can provide customer satisfaction based on the ideas of Quality First and Market Orientation in any economic environments, such as change-over to low economic growth, severe competition with other suppliers in the market, and high YEN appreciation. Stressed points of TQC activities are the following 4 items.

① Intensify management of policy to attain middle range plan.
② Intensify attaining of Q, C, D in source of process and in manufacturing process
③ Strengthen the system by which effective new products will be developed.
④ Bring up the personnel and group who can resolve management subject well.

The basic thinking of new products development is to understand current and future customer needs precisely and to supply products speedily which meet customer requirements. Stressed points of new products development activities are following 4 items.

① Sufficiency of new products planning
② Reinforcement of cooperating system with related departments
③ Positive supply of technical information to customer
④ Settlement of aimed cost and evaluation of trial products

Fig. 11 and 12 show system of new products development and cooperating system with related departments.

CONCLUSION

Several kinds of stainless steel piping have been developed for gas delivery systems in semiconductor manufacturing. These developments were done as a part of TQC-activities by taking customer needs in advance, and cooperating with our customer and Tohoku University. Authors believe firmly that these technologies will be used for many semiconductor plants and to realize non-stop production line must be the most cost-effective approach.

REFERENCES

1. H. Tomari et al : Kobe Steel Engineering Report, Vol. 41, No. 4 (1991), p. 75
2. S. Kaneko et al : The Journal of the Surface Finishing Society of Japan, Vol. 41, No. 3 (1990), p. 1
3. U.C.S. : Chemistry of Ultra-Pure Water, Realize, p. 625
4. H. Tomari et al : Kobe Steel Engineering Report, Vol. 39, No. 4 (1989), p. 57
5. T. Ohmi et al : Nikkei Microdevices, Vol. 11 (1992), p. 121

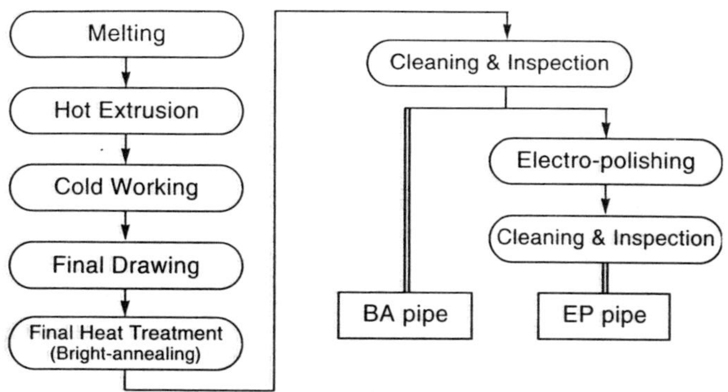

Fig. 1 Outline of manufacturing process of BA pipe and EP pipe

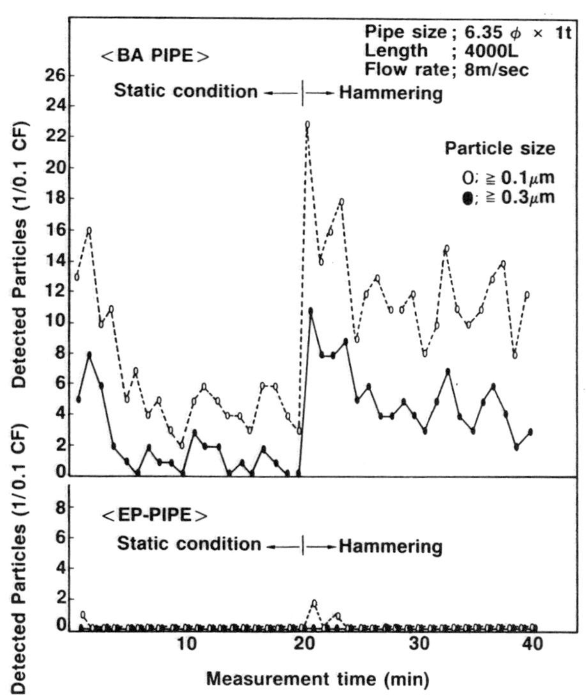

Fig. 2 Results of the particle test

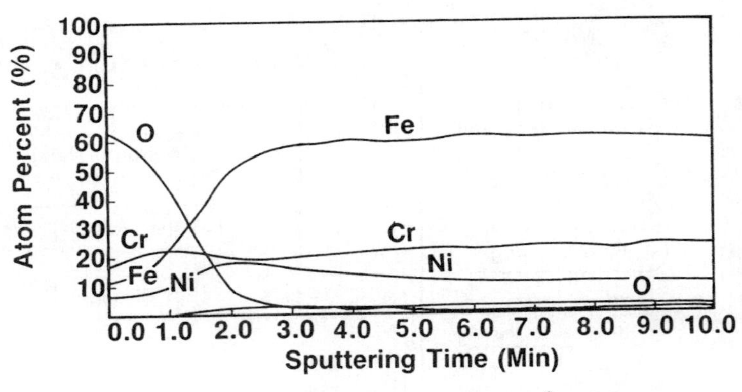

(Sputtering speed ≒ 12 Å/min)

Fig. 3 Depth-composition profile in surface film of EP-Pipe (AES)

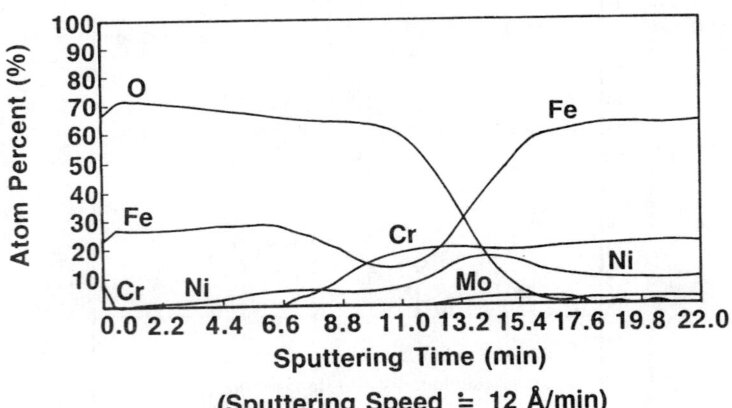

(Sputtering Speed ≒ 12 Å/min)

Fig. 4 Depth-composition profile in surface film of OP-Pipe (AES)

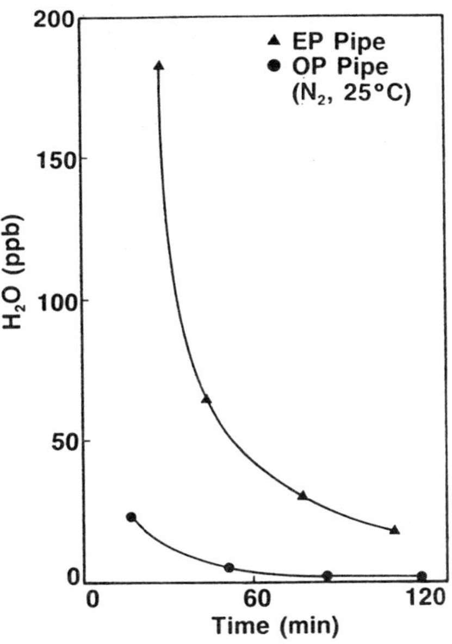

Fig. 5 Time dependence of H_2O
concentration in N_2 gas passed
through pipe at RT.

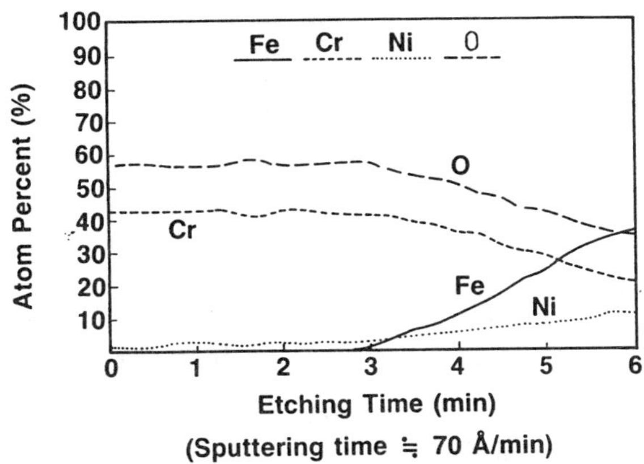

Fig. 6 Depth-composition profile in surface film of CRP-pipe (AES)

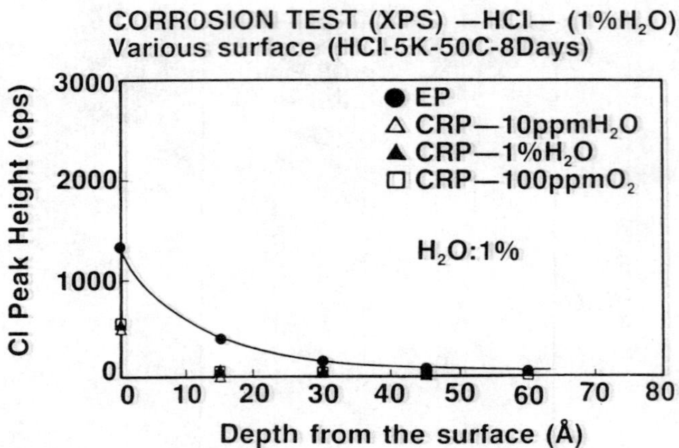

Fig. 7 Cl depth profile in surface film on various surfaces (XPS) after HCl gas exposure

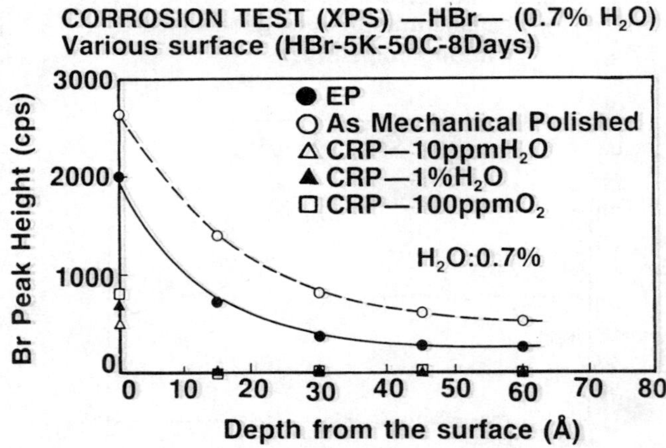

Fig. 8 Br depth profile in surface film on various surfaces (XPS) after HBr gas exposure

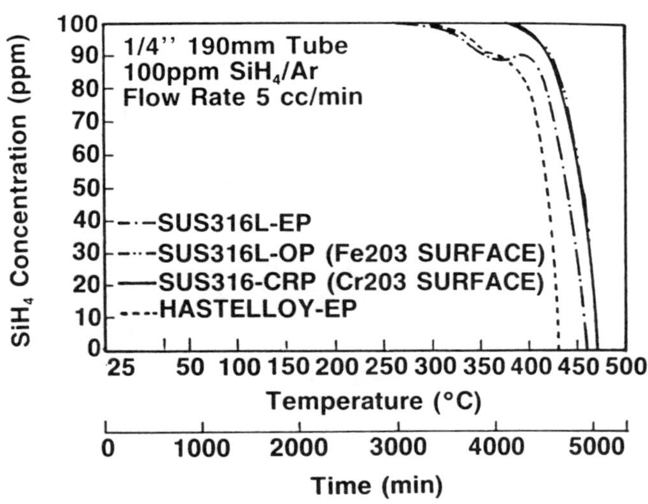

Fig. 9 Thermal decomposition of 100 ppm SiH_4 in Ar on various surfaces

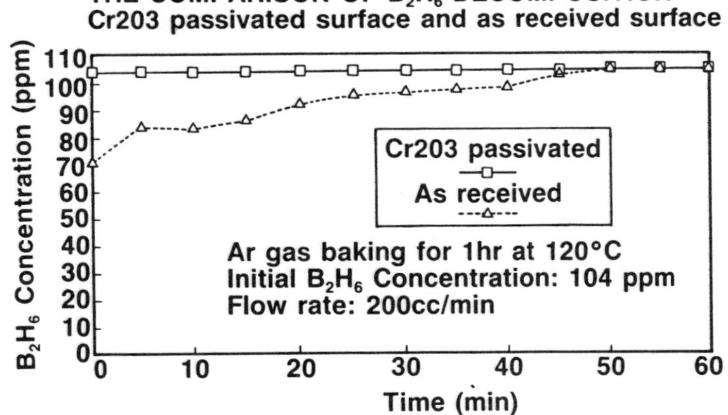

Fig. 10 Thermal decomposition of 104ppm B_2H_6 in Ar on various all-metal filter surfaces at 120°C

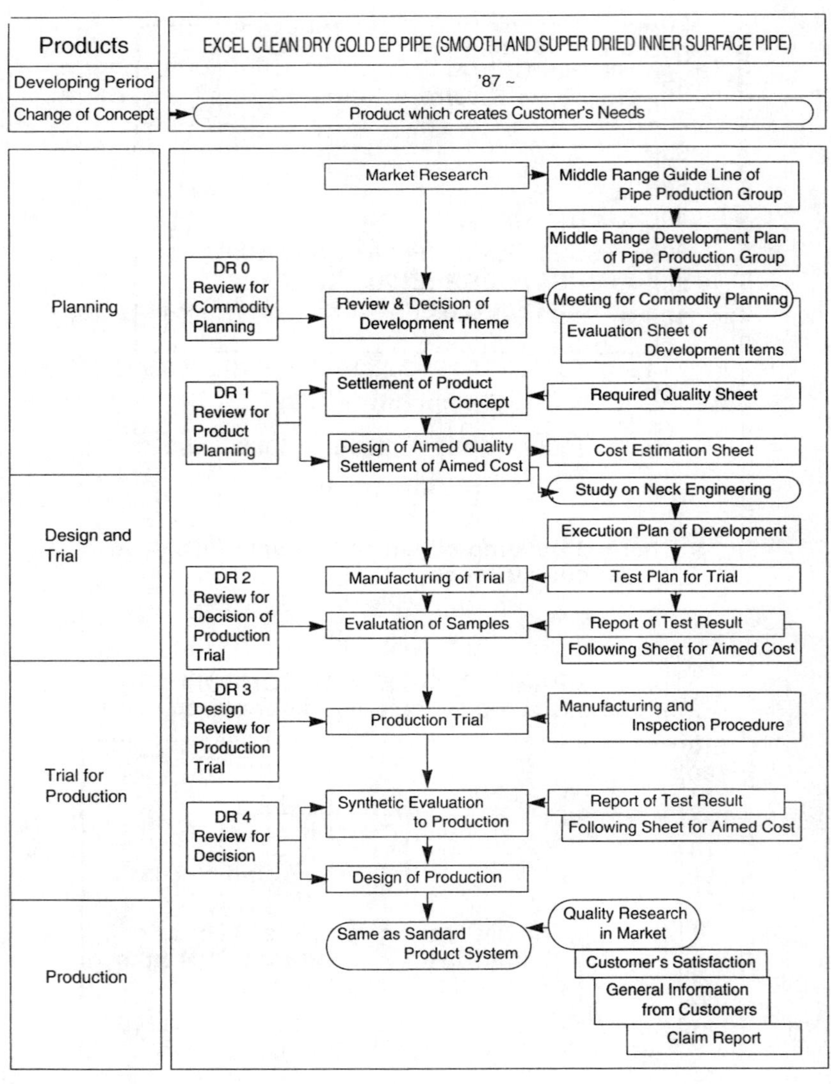

Fig. 11 Quality Assurance Activity in the Development of New Products

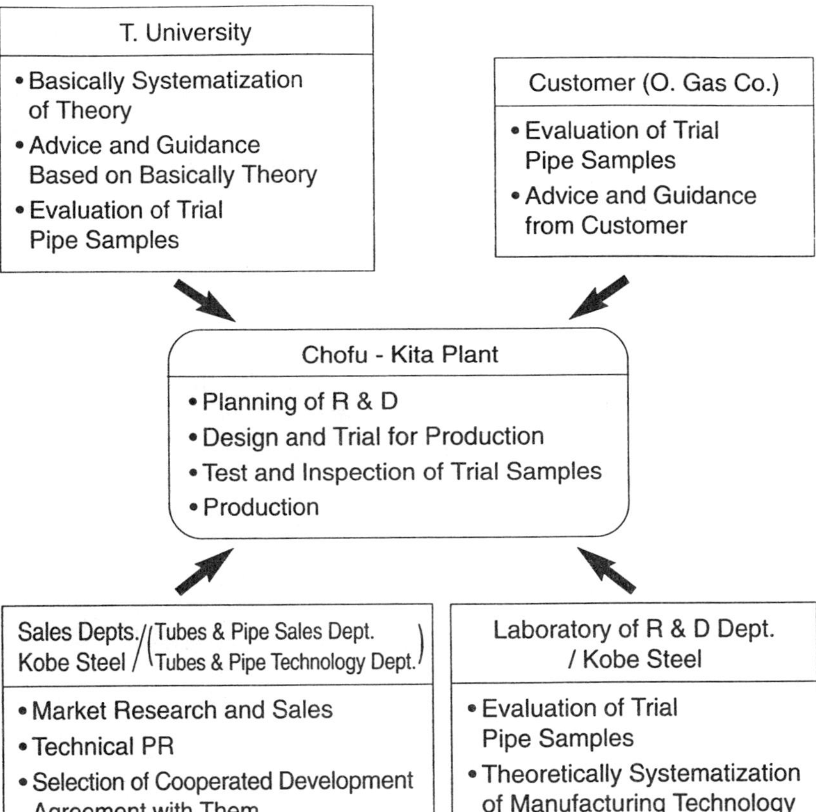

Fig. 12 Cooperating System with Related Departments

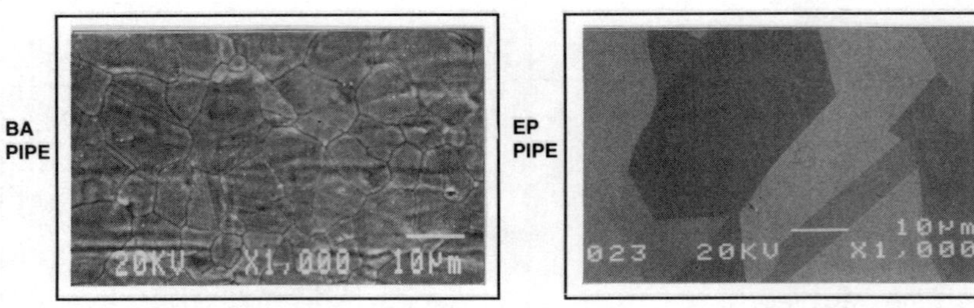

Photo 1. SEM micrograph of inner surface

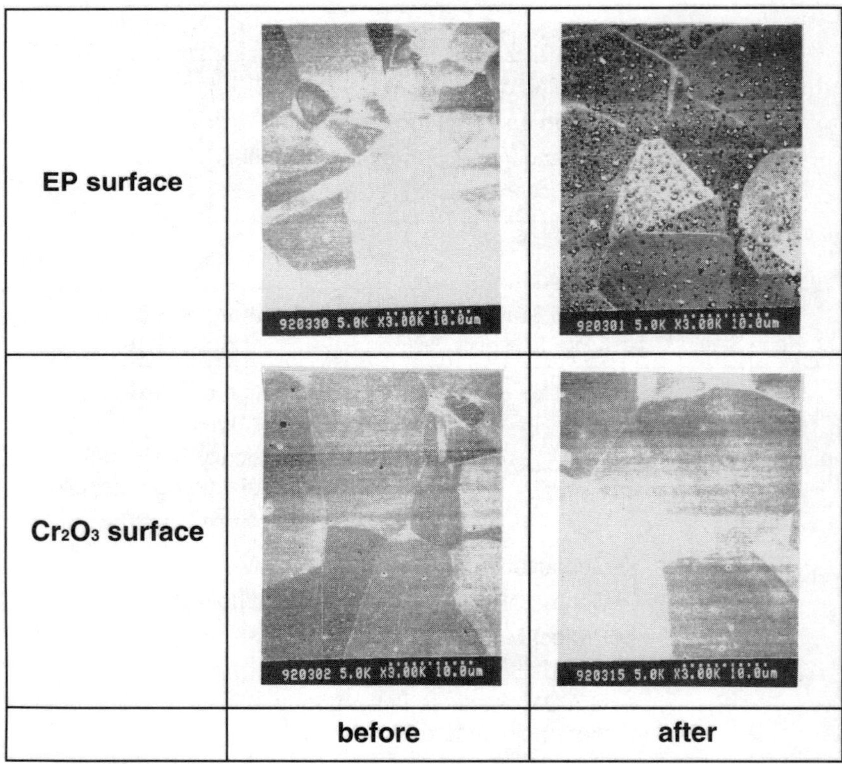

Photo 2. SEM images of electropolished and chrome oxide passivated surfaces - before and after exposure to HCl gas with about 1.5 ppm H$_2$O at 100°C for 5 days (Corrosion Test)

DEVELOPMENT OF MANUFACTURING PROCESS OF TINPLATE WITH PASSIVATION FILM CONTAINING METALLIC CHROMIUM

Fumio Aoki , Makoto Himeno, Fumio Kosumi, Shoubun Ikeda
Kawasaki Steel Corporation Chiba Works
1,Kawasaki-Cho, Chuoh-Ku, Chiba, 260 Japan

Hajime Ogata
Kawasaki Steel Corporation Technical Research Division
1,Kawasaki-Cho,Chuoh-Ku, Chiba, 260 Japan

Recently, the use of lacquered can has increased, then the lacquer wettability and lacquer adhesion has become very important. But, in the conventional tinplate, it is difficult to get good lacquer wettability and good lacquer adhesion at the same time. The tinplate with passivation film containing metallic chromium, which is formed by cathodic treatment in chromic acid solution, is superior in both properties. Kawasaki Steel Corporation revamped No.2ETL in 1991 to manufacture this high performance tinplate. The characteristic point of the process is dipping in high-concentration chromic acid solution after cathodic treatment. This treatment makes it possible to get the easy control of chromium oxide and the stable operation.

INTRODUCTION

The tinplate,which is the steel sheet coated with tin, is the popular material for food containers. Recently,in order to save the cost of cans, the tin coating weight has been decreased and the use of lacquered can has increased.Nowadays the lacquered can with less than 2.8 g/m^2 of tin is very popular then the lacquer wettability and lacquer adhesion become very important. This performance is greatly influenced by the passivation treatment in manufacturing process.

In the case of the conventional tinplate, the passivation treatment is uaually CDC(Cathodic Dichromate) treatment but it is difficult to get good lacquer adhesion and good lacquer wettability at the same time as shown in Fig.1. In order to get better lacquer adhesion, CSC (Cathodic Sodium Carbonate) treatment before CDC is sometimes adopted in Kawasaki Steel Corporation. But,CSC-CDC treatment is not good for lacquer wettability.

It is generally known that the tinplate with passivation film containing metallic chromium (M-Cr),which is formed by cathodic treatment in chromic acid solution ,is superior in both properties (1)(2)(3). In this report, we call such tinplate CSC-ECC (Cathodic Sodium Carbonate - Electrolytic Chromium Coated) type tinplate. Fig.2 shows the cross section of the each type tinplate.

CSC-ECC type tinplate is not good in solderability,but recently the use of soldered cans has decreased and the use of welded cans has increased,then the needs for CSC-ECC type tinplate has increased. And so No.2ETL in Kawasaki Steel Corporation was revamped to manufacture the CSC-ECC type tinplate.

This report describes the outline of the passivation treatment in No.2ETL for CSC-ECC type tinplate and the performances of it.

MANUFACTURING PROCESS

Fig.3 shows the general manufacturing process of tinplate in electrolytic tinning line. In 1991,we,Kawasaki Steel Corporation Chiba Works,revamped the equipment of passivation treatment in No.2ETL in order to manufacture the CSC-ECC type tinplate. Fig.4 shows the outline of passivation treatment section of No.2 ETL before and after reconstruction. This reconstruction made it possible to adopt CDC,CSC-CDC and CSC-ECC treatment in No.2ETL.

Fig.5 shows the changes of surface structure in the CSC-ECC passivation treatment. In No.1tank,the Sn-oxide ,which was formed at reflow, is removed by CSC treatment. Then, in No.3tank, the passivation layer with metallic chromium and chromium oxide is formed by cathodic treatment. And in No.4tank, the surplus chromium oxide is dissolved and removed by dipping in high concentration chromium acid solution.

The characteristic point of CSC-ECC process in No.2ETL is dipping in high-concentration chromium acid solution after cathodic treatment in chromic acid solution containing a little amount of sulfuric acid. We can easily control the weight of metallic chromium by adjusting the cathodic current in No.3tank and can control the weight of chromium oxide (Cr.OX) by adjusting the temperature and concentration of chromium acid in No.4tank.

LACQUERABILITY AND CORROSION RESISTANCE

Materials

The samples for tests were CDC type,CSC-CDC type and CSC-ECC type tinplate which were produced in No.2ETL after reconstruction. These were Al-killed T4CA, 0.2mm thick and tin coating weight was 2.8 g/m^2.In the case of CDC type and CSC-CDC type tinplate,the chromium oxide weight was about 6 mg/m^2. The metallic chromium weight of CSC-ECC type tinplate were from about 2 to 17 mg/m^2 and the chromium oxide weight was about 8 mg/m^2.

<u>Lacquerability</u>

The lacquer wettability and lacquer adhesion properties of specimens were examined by the test methods as follows.

Lacquer wettability test .

$10 \mu l$ epoxy-phenollic lacquer is dropped onto the test piece to measure the diameter of the spreading area after 24 hours at room temperature.

Lacquer adhesion test.

Glaze epoxy-phenollic lacquer($50mg/m^2$) for 10 minutes at 210 °C. Put the lattice pattern scratches with a steel pen. Stick cellophane tape and peel it off quickly to measure the peeled area of lacquer.

Fig.6 shows the lacquer wettability and Fig.7 shows lacquer adhesion of CSC-ECC type tinplate in comparison with CDC type and CSC-CDC type tinplate. It appears that CSC-ECC type tinplate with more than 2 mg/m^2 metallic chromium in passivation film had excellent wettability. It is considered that the chromium oxide film which was formed in chromium acid solution itself has good lacquer wettability. And it is found that more than 4 mg/m^2 metallic chromium is necessary to get excellent lacquer adhesion. This good lacquer adhesion of CSC-ECC type tinplate seems to be caused by strong bonds of Sn / M-Cr / Cr-OX.

<u>Corrosion resistance</u>

In lacquered cans, one of the typical corrosion is underfilm corrosion with blistering which is commonly observed in the head space area of the can. To simulate this type of corrosion, the blister test was established. Test method is as followed.

The blister test.

The specimen on which epoxy-phenollic lacquer (50 mg/m^2) is glazed for 10 minutes at 210°C is half immersed in boiled tomato juice. The juice container is sealed tightly and kept at 55°C for 7days. The blister formation on the specimen is evaluated.

Fig.8 shows the blister formation of the CSC-ECC type tinplate in comparison with CDC type and CSC-CDC type tinplate .It appears that the blister formation became excellent with increase of metallic chromium coating weight and there was no blisters over 7 mg/m^2 of metallic chromium. It is considered that the good corrosion resistance was caused by the strong lacquer adhesion.

CONCLUSIONS

1.   The passivation treatment section of No.2ETL in Chiba Works was revamped in 1991 to manufacture the CSC-ECC type tinplate,then has been operating smoothly.

2.   It is reconfirmed that the CSC-ECC type tinplate has good lacquer wettability and good lacquer adhesion at the same time in comparison with CDC type or CSC-CDC type tinplate and also good corrosion resistance.

3.   Now we have been applying the CSC-ECC type tinplate to more commercial use of lacquered cans and have succeeded to meet the demands of customers for tinplate.

REFERENCES

Journals

1. K.Shimizu: Camp-Rep. of Iron and Steel Inst.Japan,1988,1,720
2. Y.Takei, O.Yoshioka: Tetsu to Hagane,1973, A139

Society Softbound Proceedings Series

3. H.Nakakouji: 4th International Tinplate Conf.London,1988,163-176

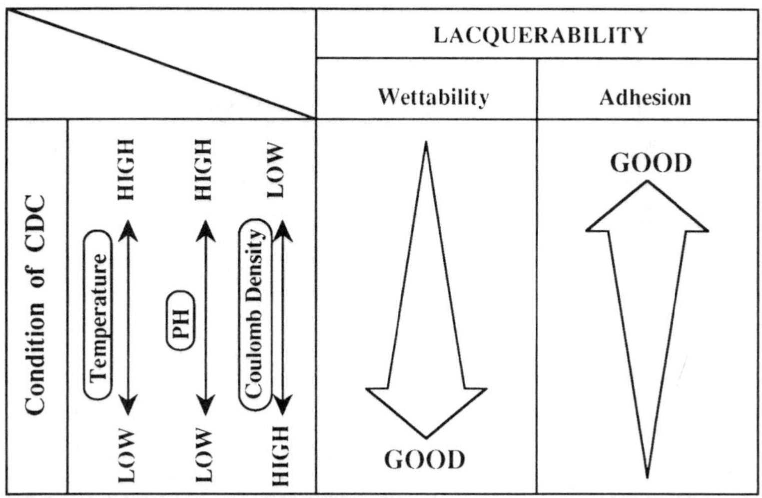

Fig. 1 Relationship between Lacquerability and the Condition of CDC

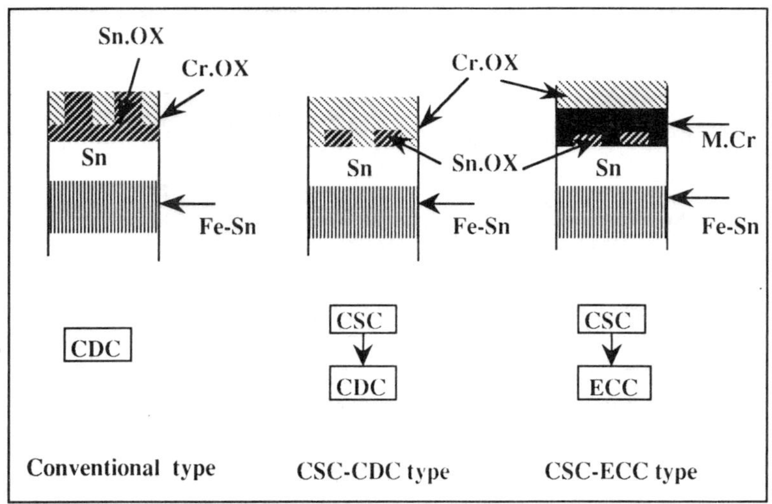

Fig. 2 Cross Section of Tinplate

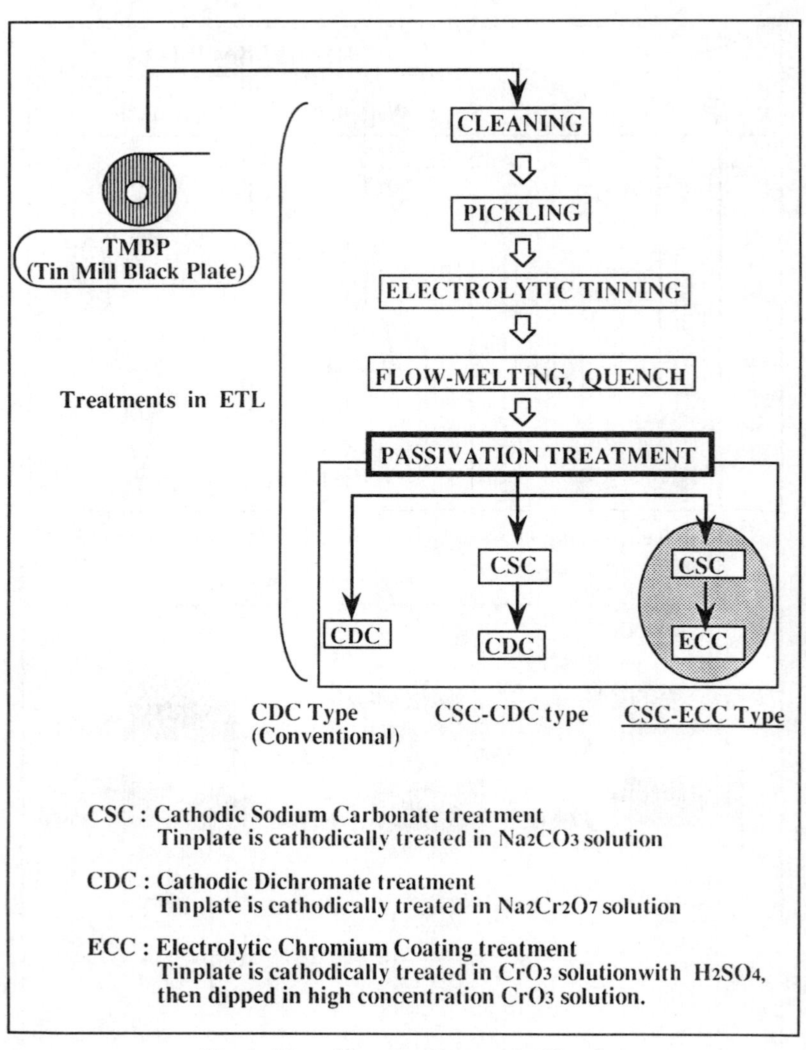

CSC : Cathodic Sodium Carbonate treatment
    Tinplate is cathodically treated in Na_2CO_3 solution

CDC : Cathodic Dichromate treatment
    Tinplate is cathodically treated in $Na_2Cr_2O_7$ solution

ECC : Electrolytic Chromium Coating treatment
    Tinplate is cathodically treated in CrO_3 solutionwith H_2SO_4,
    then dipped in high concentration CrO_3 solution.

Fig. 3   Manufacturing Process forTinplate

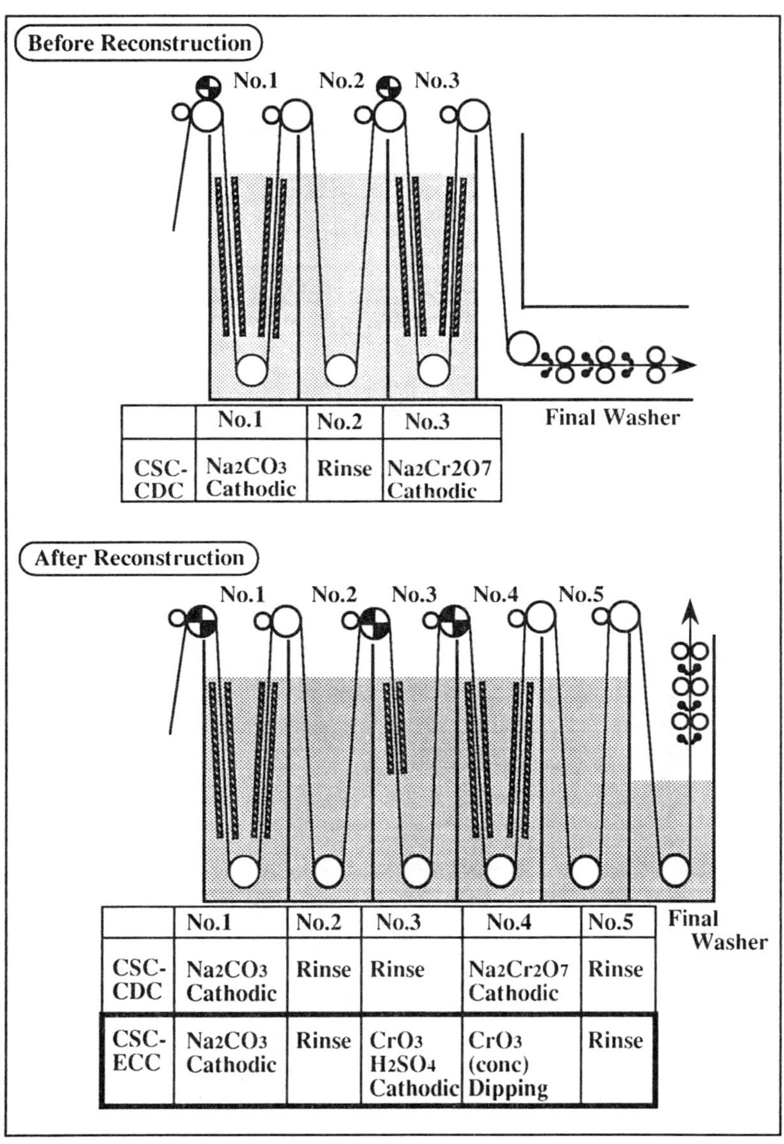

CSC-CDC	No.1	No.2	No.3
	Na_2CO_3 Cathodic	Rinse	$Na_2Cr_2O_7$ Cathodic

	No.1	No.2	No.3	No.4	No.5
CSC-CDC	Na_2CO_3 Cathodic	Rinse	Rinse	$Na_2Cr_2O_7$ Cathodic	Rinse
CSC-ECC	Na_2CO_3 Cathodic	Rinse	CrO_3 H_2SO_4 Cathodic	CrO_3 (conc) Dipping	Rinse

Fig.4   Equipment of Passivation Treatment in No.2ETL

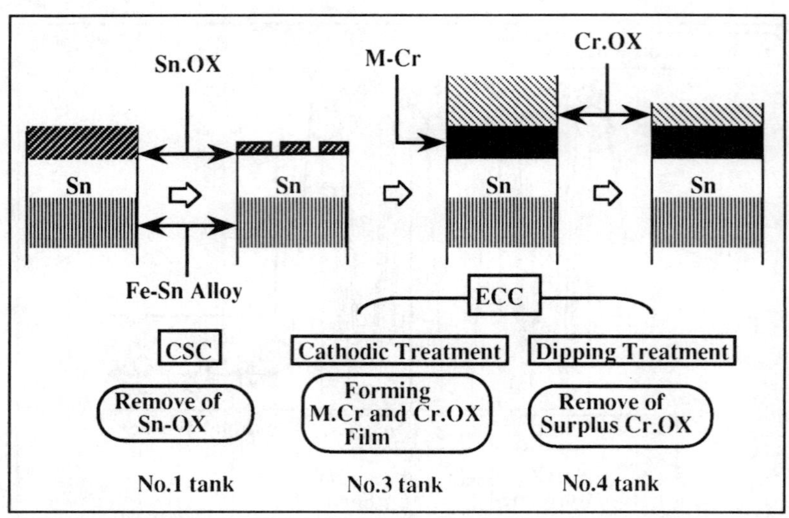

Fig.5 Changes of the surface structure in CSC-ECC treatment

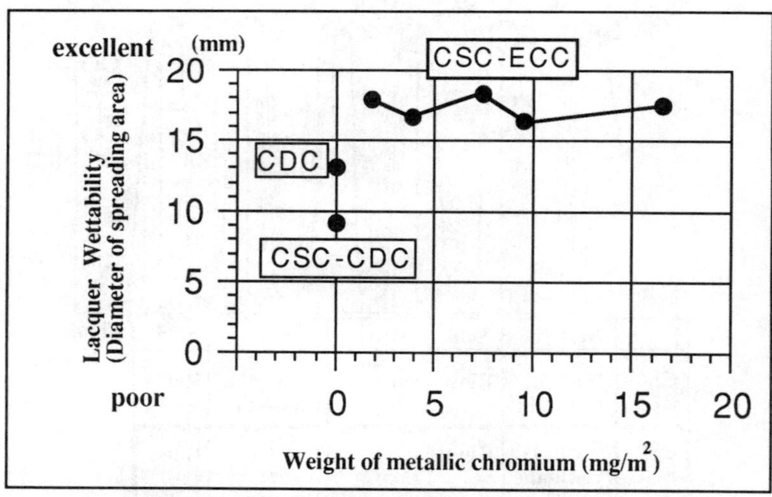

Fig.6 Relationship between weight of metallic Cr
and lacquer wettability

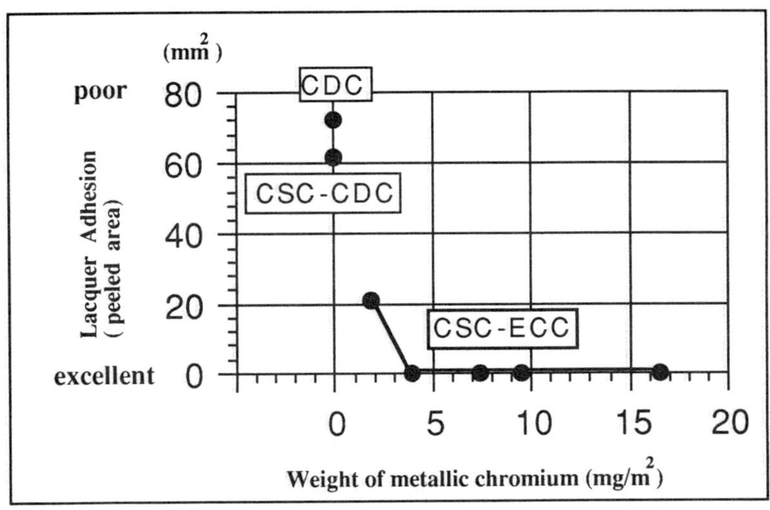

Fig.7 Relationship between weight of metallic Cr
and lacquer adhesion

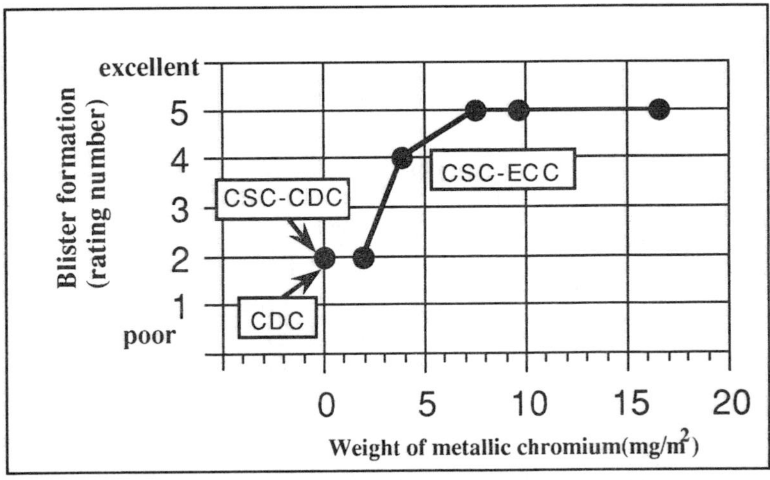

Fig.8 Effect of metallic Cr weight on blister formation

POWER COST AND CATHODE QUALITY OF THE CELLHOUSE
BASED ON NEW CONCEPTS
AT MITSUBISHI AKITA ZINC REFINERY

A.Katai
K.Sato
A.Kakimoto
Mitsubishi Materials Corporation
Akita Refinery
3-1-18 Barajima Akita Japan 010

ABSTRACT

The unit cost of electricity in Japan is
extremely high. Much effort has been made to
reduce the cost of electricity in zinc
refineries. At Mitsubishi Materials Akita
Zinc Refinery, the cellhouse based on new
concepts was developed and is in operation.
During the actual operation, we could obtain
the estimated results concerning the
electric power consumption and the new
information concerning the behavior of
lead in deposited zinc. In this paper, actual
operation data of the new cellhouse are
discussed and compared with the conventional
cellhouse.

INTRODUCTION

The unit cost of electricity in Japan has risen
rapidly since 1974 when the oil crisis occurred. Today,
the power companies have introduced a price system in
which electric power cost varies according to the season
and the time of day. So we have to change the load of
electrolysis to utilize the low price times from 10 pm to
8 am and holidays including Sundays.

Under these circumstances various improvements had
been made to reduce the power cost in electrolysis. In
1989, a cellhouse of zinc electrowinning based on new
concepts was constructed on a commercial scale and was
put into operation.

The major modification of the cellhouse is that a
kind of insulating "frame" is attached to the anode.
Anodes and cathodes are put together with these frames to
make a unit and those electrodes are treated
simultaneously for stripping and cleaning. This system

makes it possible to remove the anode crust every two days and to reduce electrodes distance, which lead to the reduction of electric power consumption.

In conventional cellhouse, at Mitsubishi Akita Refinery, electrolysis is stopped after a 26 day cycle. The anodes are pulled out of the cells to remove the anode crust. So it can be possible to observe the surface of the anodes only after a 26 day cycle. With the new method, the surface of the anodes are observed every 48 hours to attain the new information about the short circuits between the electrodes. This new method decrease short circuits. This causes a decrease in lead contamination which increase the quality of deposited zinc.

During the actual operation, the distance between electrodes was extended 20% over the initially designed distance as the electrode handling machine was improved in 1991. In this paper, the actual operation data of electric power consumption and cathode quality are discussed and compared with those of a conventional cellhouse.

ELECTRIC POWER CONSUMPTION IN ELECTROLYSIS

Fig.1 shows the zinc production at Mitsubishi Akita Zinc Refinery. At the beginning of the 1970's, zinc production was at a maximum with full capacity of silicon rectifier. In 1973 the unit cost of electricity rose in accordance with the sudden rising of the crude oil prices. Therefore, this caused production levels to decrease by 70%. In 1976, the power companies introduced a price system in which the electric power cost varied according to the season and the time of day. We were obliged to control the load of electrolysis during nighttime and daytime.

Fig.2 shows the electric power consumption per unit output. In the latter half of 1970's, the energy use was 3400-3500 KWH/T-Cathode at full operation. At the end of 1970's, various improvements were made to maintain the electric power consumption of 3150-3200 KWH/T-Cathode. Those improvements were as follows:

```
1)A rise in temperature and free acid in electrolyte
     F.A.    115-120 g/1    →      180 g/1
     Temp.   38-39 °C       →    39-41 °C
2)Removal of Mg in electrolyte
     Mg          12 g/1      →      7.2 g/1
```

      3)Shortening the cycle of anode cleaning
                28-30 days →   24-26 days

     Fig.3 shows load change of electrolysis at our
refinery. It was necessary to save costs by utilizing low
price power during the specified times and minimizing
high price power during the daytime. However the
operation with the full capacity of silicon rectifier
made it impossible to utilize the electric power
during the nighttime. Under these circumstances, it was
planned to raise the nighttime ratio by transferring the
silicon rectifier from Hosokura Zinc Refinery to
the Akita Zinc Refinery and by increasing the number of
cells. A new electrowinning system for zinc was adopted,
which is completely different from the conventional
concept. After a year and a half of testing,
the commercial base cellhouse has been operating since
1989.

OPERATION IN THE NEW CELLHOUSE

     The fundamental concepts of the new cellhouse for
reducing the electric power consumption are as follows.
The major factors that determine the electric power
consumption are cell voltage, current efficiency and
current converting efficiency of the silicon rectifiers
as the formula [1] shows.

$$W=E/(1.22 \times \eta_1 \times \eta_2) \times 10^7 \qquad -------- \quad [1]$$

     W:electric power consumption (AC KWH/T-Cathode Zn)
     E:cell voltage (V)
    η_1:current efficiency (%)
    η_2:current converting efficiency (%)
  1.22:electrochemical equivalent for zinc per 1 AH(gZn)

It is also known that the cell voltage E is calculated
from the formula [2]. (1)

$$E=2.67+\beta i \quad ------------[2]$$

   $\beta:\beta_1+\beta_2+\beta_3$
    $\beta_1=\ell/k$
   ℓ:inter electrode distance (face to face) (cm)
   k:specific electric conductivity of
      electrolyte ($\Omega^{-1}cm^{-1}$)
   β_2:coefficient for anode crust
   β_3:coefficient for other cause
   i:current density (A/cm^2)

As predicted by formula [2], one of the effective measures to reduce the cell voltage, and hence, electric power consumption, is to reduce the distance between electrodes. On the other hand, the resistance caused by the anode crust increases the cell voltage E. Therefore, the other effective measure to reduce the cell voltage is to remove the anode crust frequently. A new cellhouse was planned by introducing these concepts. A kind of insulating "frame", shown in Fig.4, was attached to anodes in order to maintain a short and accurate distance between electrodes. Fifteen cathodes and sixteen anodes were put together with these frames to make a unit. Two units are submerged in a cell. (Fig.5) An entire electrode unit is pulled out of the cell for treatment. During this process, zinc is stripped from the cathodes and the anodes are cleaned at the same time. The anode crust is removed every 48 hours. This method makes it possible to reduce the resistance β_2 caused by the anode crust.

Commercial scale operation based on these new concepts have been in operation since March 1989. However, the electrode handling machine which was called EHM was not working efficiently. The designed treatment time of 6 minutes/cycle could not be achieved. The EHM was reconstructed in March 1990 and the following are the new innovations. The EHM was installed over the cells and handles a series of operations in a fully automated manner, such as pulling the plates out, handling electrodes through the washer and stripping machine, and starting the electrolysis. The EHM was divided into three machines identified as No.1 EHM, No.2 EHM and ET (electrodes table). The No.2 EHM pulls the unit out of the cells, conveys the unit to ET and inserts the unit which is treated by the No.1 EHM into the cell. The No.1 EHM is equipped with 27 pairs of hangers that pick up the hooks of 13 cathodes and 14 anodes. These 27 pairs of sliding hangers travel in the direction of the width of No.1 EHM. This is called "opening the unit". When the unit is completely disassembled, the pitch between each plates is 160 mm. After opening the unit, No.1 EHM travels through the electrode cleaning machine and puts the cathodes and anodes into the stripping machine. Once the process is completed, the No.1 EHM reverses and moves into the wet type polishing machine. After polishing the unit is closed and No.1 EHM sets the reassembled unit on ET. These mechanical improvements can achieve the treating time of 6 minutes/cycle. On the contrary, in the case of longer deposition times due to mechanical troubles, short circuits were increased. This caused the increase of lead content of deposited zinc and decreases in current efficiency.

During the actual operation, removal of the anode crust incompletely caused the scale-like crust which created short circuits. In order to continue the stable operation for a long time under the status quo of the anode material and anode treatment, we were obliged to take a step backwards from the shortest distance between electrodes of (14.5mm) in the world. In October 1991, by constructing a unit with 13 cathodes and 14 anodes, the distance between electrodes is now 17.5 mm. Also, at this time the sliding hangers of No.1 EHM was reconstructed.

ESTIMATED ELECTRIC POWER CONSUMPTION OF NEW CELLHOUSE

Table I shows a comparison between a new cellhouse and a conventional cellhouse before and after reconstruction of the new cellhouse.

Table I   Calculated data of new cellhouse

	Conventional	New Concept (1989)	New Operation (1991)
current (A)	15,200	15,200	15,200
current density (A/m^2)	507	507	507
current efficiency (%)	89.31	89.31	89.31
cell voltage (V)	3.431	3.158	3.192
current converting efficiency of Si-R (%)	96.62	96.62	96.62
electric power consumption (KWH/T·cathode Zn)	3,302	3,000	3,032

As predicted by formula [2], reduction of the distance between electrodes and the resistance of the anode crust makes β, and hence, cell voltage E, lower than that of the conventional cell. It was calculated that there would be an improvement of the electric power consumption by 300 KWH/T-Cathode. Even after reconstruction which increased the distance between electrodes in 1991, it was predicted that there would be an improvement of 270 KWH/T-Cathode.

During the actual operation of the new cellhouse, the maximum capacity of the silicon rectifier which is

18 KA is used. Under this condition, the data shown in table II was obtained. In this table, current converting efficiency was worse because of an outdated silicon rectifier. By reducing the number of anodes and cathodes for reconstruction, this increased the current density which caused the decrease of current efficiency. The result was 3294 KWH/T-Cathode in electric power consumption.

Table II Actual operation data of the new cellhouse

	Conventional	New Concept (1989)	New Operation (1991)
distance between center of each cathode (mm)	73.0	41.0	47.0
inter electrode distance (face to face) (mm)	30.5	14.5	17.5
number of cathode per a cell	24	30	26
number of anode per a cell	25	32	28
current (A)	15,200	18,000	18,000
current density (A/m^2)	507	480	554
β (Ωcm^2)	15.0	9.62	10.29
current efficiency (calculated) (%)	88.15	89.31	85.65
cell voltage (V)	3.431	3.132	3.240
current converting efficiency of Si-R (%)	96.62	94.13	94.13
electric power consumption (KWH/T·cathode Zn)	3,302	3,054	3,294

ELECTRIC POWER CONSUMPTION AND CATHODE QUALITY

Deposition Time vs Current Efficiency

Fig.6 shows the relationship between the deposition time and current efficiency of the new cellhouse before its reconstruction in 1991. Fig.7 shows the same relationship in the conventional cellhouse. Here, the deposition time is defined as the converted time of deposition. (Total amount of current(KAHr) is divided by 18 KA.) In the conventional cellhouse, current efficiency was almost constant, independent of deposition time. On

the contrary, in the new cellhouse, current efficiency
depends on the deposition time. It is thought that, in
the new model, the deposited zinc and the anode crust
makes the distance between electrodes shorter than the
initial distance of 14.5 mm. Therefore short circuits
easily occurred.

Fig.8 shows the relationship between the deposition
time and the current efficiency of the new cellhouse
after reconstruction in 1991. Compared with Fig.6, the
ratio is lower than before reconstruction even though
it has a larger dispersion.

Short Circuits vs Current Efficiency

Fig.9 shows the relationship between current
efficiency and the number of short circuits per cell in
the new cellhouse. In Fig.9 also shows that current
efficiency is related to the number of short circuits. As
the distance between electrodes in the new model are too
short compared with that of the conventional one, short
circuits are caused easily by the anode crust which is
dropped from the anode surface. This figure shows that it
is extremely crucial to control the number of short
circuits below 1.5 per cell in order to achieve the
current efficiency above 85% during actual operation.

To reduce the anode crust, as mentioned above,
the anode crust must be cleaned off with pressurized
water every 48 hours. At the time of reconstruction in
1991, the pressure of water was raised to remove the
anode crust more efficiently.

Pb vs Current Efficiency and Short Circuits

Fig.10 shows the relationship between the current
efficiency and lead content in deposited zinc. Fig.11
shows the relationship between the lead content in
deposited zinc and the number of short circuits. The
above data was obtained during the addition of a Pb
controlling agent of 1.5-2.0 Kg/T-Cathode. It is
generally said that the dissolved lead-silver anodes
caused by short circuits create high lead content in
deposited zinc. As the number of short circuits
increases, the lead content of deposited zinc increase.

Fig.12 shows the movement of the lead content in
deposited zinc in recent years. Before June of 1992,
mechanical troubles made the deposition time longer than
48 hours, so that caused a lowering of current
efficiency. This was the reason why the lead content

in deposited zinc was high in the new cellhouse. After solving the mechanical troubles, the lead content in the new cellhouse has been on the same level as that in the conventional cellhouse.

Current Density vs Cell Voltage

Fig.13 shows the relationship between the current density and the cell voltage in the new cellhouse. Cell voltage E is calculated by the equation [3] in the new cellhouse.

$$E=2.67+10.29i \ \text{-------}[3]$$

From the obtained data during the actual operation, the equation $E=2.68+8.45i$ was obtained. Total coefficient for voltage drop β is smaller than the calculated β.

Table III   Actual Operation Data

	July	August	September
new cell			
current efficiency (%)			
observed	86.80	86.62	85.18
(calculated)	(87.96)	(87.91)	(87.34)
electric power consumption (KWH/T·cathode Zn) observed	3,092	3,083	3,155
(calculated)	(3,055)	(3,071)	(3,088)
average current density (A/m²)	422	428	435
conventional cell			
current efficiency (%)			
observed	86.50	91.54	91.07
(calculated)	(87.90)	(87.91)	(88.09)
electric power consumption (KWH/T·cathode Zn) observed	3,138	3,039	3,079
(calculated)	(3,042)	(3,066)	(3,074)
average current density (A/m²)	314	331	341

Fig.14 shows the relationship between the current density and the cell voltage in the conventional cellhouse. In this case, E is calculated by equation[4].

$$E=2.67+15.0i ------[4]$$

From the observed operational data in the conventional cellhouse, the equation $E=2.71+12.1i$ is obtained. This tendency of the observed ratio being smaller than the calculated ratio is similar to the relationship between the new model's calculated and observed ratios.

Electric Power Consumption

Actual operational data of the current efficiency and electric power consumption is shown in Table III. It is shown that the estimated results have been obtained in the new cellhouse.

SUMMARY

1) After reconstruction in which the electrode distance was extended in 1992, the predicted electric power consumption was obtained during actual operation.

2) The electrode handling system made it possible to observe the anode condition every 48 hours. The new information about the anode crust and short circuits was obtained by this new process.

3) It has been shown that lead contamination in deposited zinc depends on the condition of the anode crust and deposition time.

We developed the new type cellhouse by handling the anodes and the cathodes simultaneously and by reducing the distance between the electrodes. During the actual operation, the estimated results of the quality and electric power consumption have been obtained. In order to reduce the cost further, it will be necessary to improve the frames, the anode treatment.

REFERENCES

Books

1.M.Takahashi,N.Masuko : Chemistry of Industrial Electrolytic Processes AGNE , 1979.

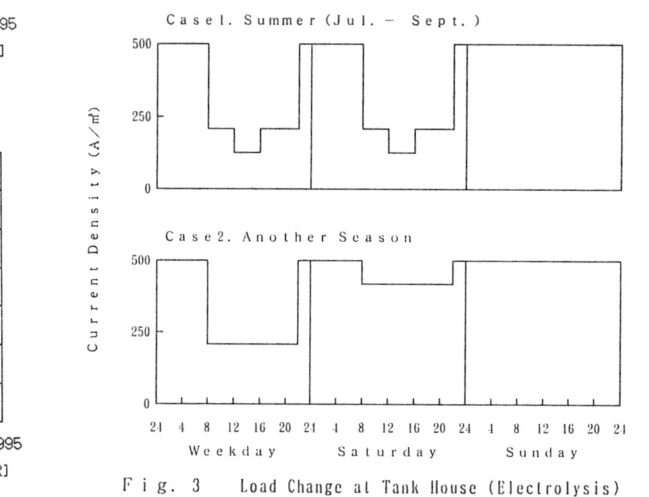

Fig.1   ZN PRODUCTION AT AKITA REFINARY

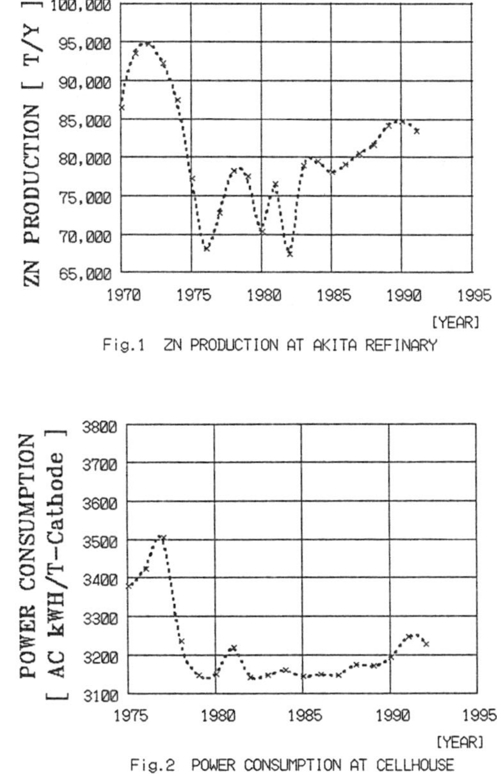

Fig.2   POWER CONSUMPTION AT CELLHOUSE

Fig. 3   Load Change at Tank House (Electrolysis)

145

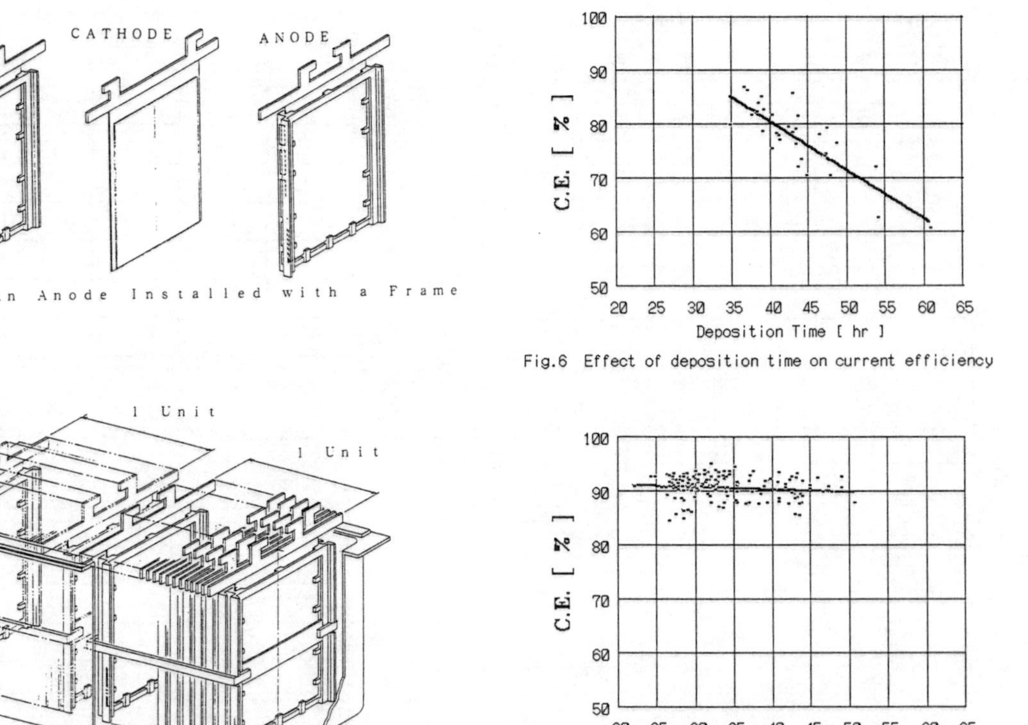

Fig. 4   An Anode Installed with a Frame

Fig. 5
Two Electrode Units
Submerged in a Cell

Fig.6 Effect of deposition time on current efficiency

Fig.7 Effect of deposition time on C.E. (Conventional Type)

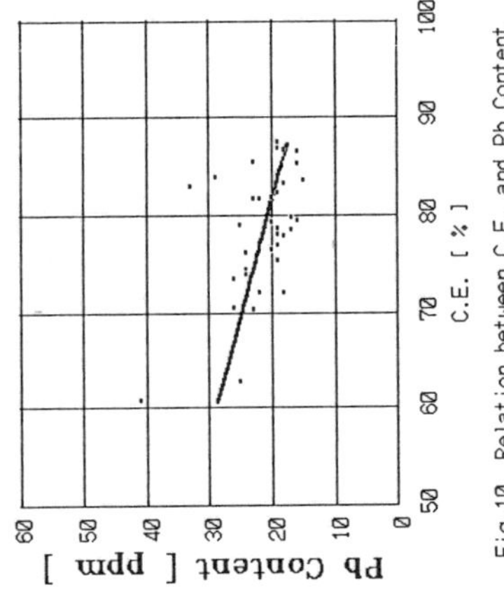

Fig.8 Effect of deposition time on C.E. (New Type)

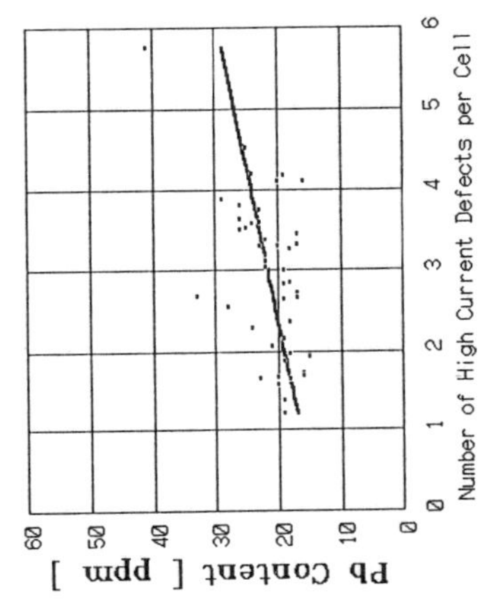

Fig.10 Relation between C.E. and Pb Content

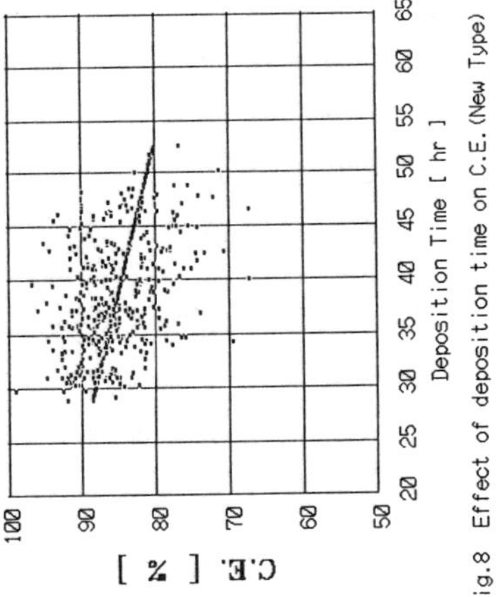

Fig.9 Relation between Short Circuits and C.E.

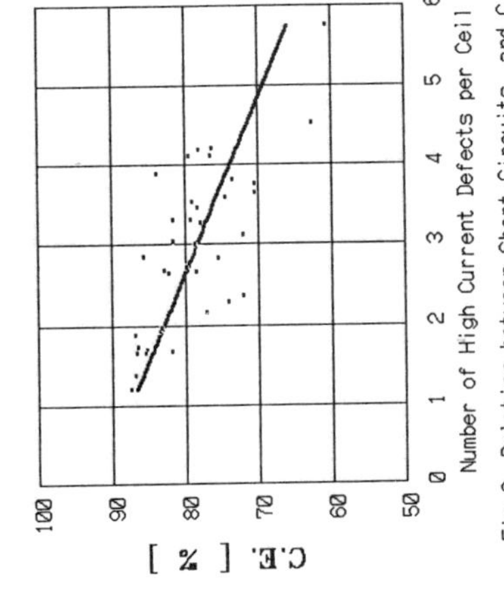

Fig.11 Relation between Short Circuits and Pb Content

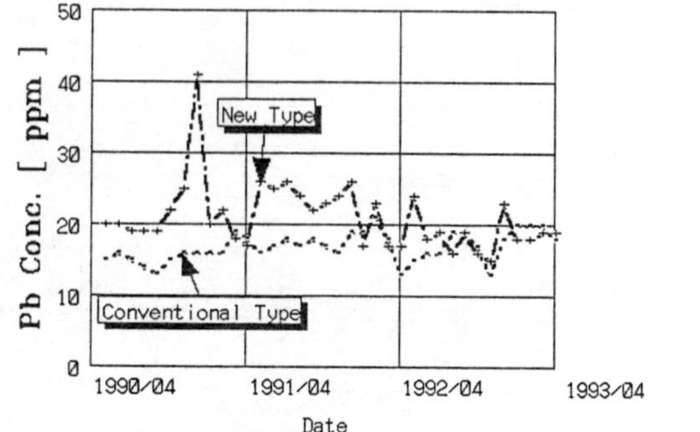

Fig.12 Behavior of Pb conc. in zinc cathode

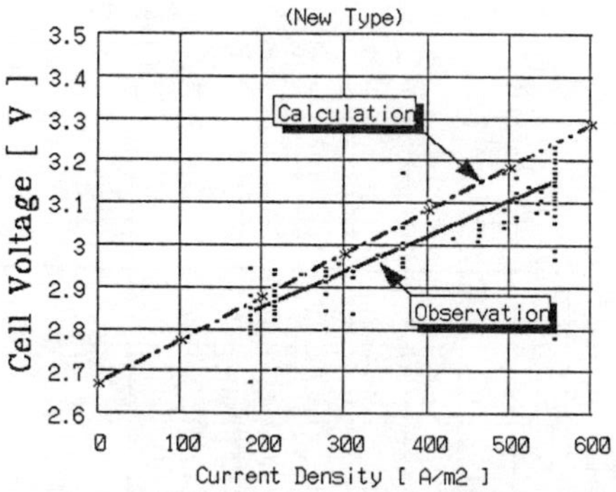

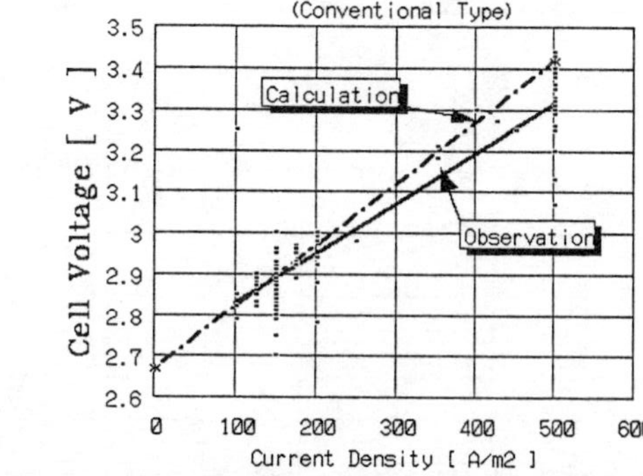

Fig.13 Relation between current density and cell voltage

Fig.14 Relation between current density and cell voltage

148

QUALITY CONTROL OF ELECTROPLATED
ALLOY STEEL SHEETS

<u>Tetsuaki Tsuda</u>*, Masanari Kimoto**, Junichi Uchida**,
<u>Kazuo Asano</u>***, Atsuyoshi Shibuya**, Yoshihiko Hoboh*,
and Ryoichi Nomi****

Sumitomo Metal Industries, Ltd.
*   Wakayama Steel Works, 1850 Minato, Wakayama-city 640 Japan
** Research and Development Div., 1-8 Fuso-cho, Amagasaki-city 660 Japan
*** Kashima Steel Works, 3 Oaza Hikari, Kashima, Ibaraki, 314 Japan
**** Tokyo Head Office, 1-1-3 Ohtemachim Chiyoda-ku, Tokyo 100 Japan

ABSTRACT

Electrogalvanized steel sheets coated with zinc-based alloy, such as Zn-Ni, Zn-Fe, Zn-Co and etc., are utilized for automotive application to meet or exceed standards requiring protection against cosmetic corrosion for 5 years and steel perforation corrosion for 10 years. Anomalous codeposition features very largely in these alloy electroplating, where current density and mass-transfer of cationic species are critical factors affecting alloy composition, in addition to bulk electrolyte composition. Precise control of electroplating parameter is essential for assuring uniform alloy plated steel products. Integrated controls of process variables monitored by high-morale workers and supported by well-kept preventive maintenance, are the key for the successful production of prime-quality electrogalvanized steel.

INTRODUCTION

A wide variety of zinc-based alloy electroplated steels have been applied in automotive body skin panels to secure long-term warranties for rust prevention, and these warranties have recently been extended to chassis and/or structural parts. Electrodeposition of zinc alloy, in particular, with iron-family elements shows anomalous codeposition characteristics, which are significantly affected by mass-transfer of cations in electrolyte(1,2). Process variables should be precisely kept within narrow controlling-windows to obtain optimum alloy content and coating mass of electroplated coatings. Various methods of electrolyte flow injection have been attempted to avoid powdery deposits caused by limited mass-transfer of cations to the surface of the cathode strip at high current densities. Velocity profiles in the vicinity of a moving-strip surface for commercial electroplating cells may be characterized as a modified Couette-flow with the superposition of injected flows. These deviate from typical velocity distributions of parallel turbulent-flow through a rectangular channel and/or Couette-flow between two flat walls, where one is in motion and the other fixed. As may be seen in Fig. 1, counter-flow injection of electrolyte towards a moving-strip surface is considered to be one of the most effective methods to provide both uniform and high mass-transfer rate of electro-chemical species, regardless of the types of electroplating cells adopted, such as vertical-cells, horizontal-cells and radial-cells(3,4,5,6,7). The incident angle of jet-flows to the traveling pass of a moving-strip is of great interest for the design of nozzles in view of uniform mass-transfer. In this study Zn-Ni and Zn-Fe alloy plating was carried out in a rectangular flow-channel cell with inclined fluid injection for the purpose of examining the effect of wall-jet flow. This was intended to simulate the flow conditions of a practical injection cell. Simultaneous co-evolution of hydrogen bubbles during anomalous alloy deposition seemed to play a major role in alloy plating at high current densities not only through raising pH near the cathode but also by enhancing mass-transfer in the diffusion boundary layer on the cathode surface(3,4,8). Winand et al.(9,10) proposed an experimental technique to investigate mass-transfer rate at a cathode by the codeposition of trace cadmium ions competing with the electrodeposition of zinc. In this study codeposition rates of trace metallic ions were measured to envisage mass-transfer at the cathode during Zn-Ni and Zn-Fe alloy-plating. Everlasting quality control activities towards the customers' satisfication are evolutionary processes beginning at the research & development phase through the mass-production phase, as indicated in Table 1. Laboratory scale developments for optimum cell design are focused in this paper.

EXPERIMENTAL

Electrodeposition was performed on a piece of steel sheet in an acidic solution containing zinc sulphate, nickel sulphate or ferrous sulphate and supporting electrolytes under galvanostatic conditions. Streamwise variation of alloy composition was investigated through chemical analysis, involving inductively coupled plasma atomic emission spectrometry. Crystal phase/structure of alloy deposits was investigated by X-ray diffractometry. Concentrations of trace metal ions were mostly measured by means of differential pulse anodic stripping voltammetry at a hanging mercury drop electrode.

Figure 2 illustrates a schematic diagram of a cell built to allow a turbulent jet of electrolyte to the cathode strip at 0 degree and 45 degree angles. The leading-edge of cathode was elongated upstream from that of the anode by 5 cm to reduce any nonuniformity of current distribution near edges of parallel plate electrode system. The cell was made of transparent polyvinyl-chloride with a 1.5 x 3.5 cm rectangular cross section (d_e = 2.1 cm). Electrolyte was introduced to either a parallel-flow inlet or to an oblique angle inlet to compare the effect of a wall-jet flow with that of the well-defined parallel flow. The anodes were pure nickel thin sheet or cold rolled steel sheet and cathode was a thin sheet of Al-killed low carbon steel. Trace metal ions were added to the electrolyte in the form of their soluble salts. The range of average Reynolds number covered was 5,400 - 50,000.

RESULTS and DISCUSSION

The results of Zn-Ni electrodeposition in rectangular flow-channel cells with varying anode-to strip gaps and solution flow-rates clearly show that Ni content in deposits is a function of Reynolds number under given plating operation conditions. The representative length is equal to the hydrodynamic equivalent diameter calculated from the anode-to-strip gap, as is seen in Fig. 3. Of particular significance for hydrodynamic effects on the Ni content in alloy-coating is the fact that the Ni content is dependent on the anode-to-strip gap even for identical mean flow velocity.

It is clearly seen in Fig. 5 that there is a significant difference between wall-jet flow and parallel flow in relation to the streamwise uniformity of alloy deposits. For a parallel flow system, neither Ni content in the alloy coating nor overall rate constant of cadmium codeposition varied with location along the cathode. It is well known that the mass-transfer entry-length in turbulent flow is very short(11), even when flow separation occurs at a sharp-edges inlet(12) or there is an obstacle inside a tube(13), which is consistent with the present results. As compared to the parallel turbulent flow system, it would be expected that the mass-transfer rate should depend on the distance from the incident point for wall-jet systems(13,14), where a jet of electrolyte is issued from a nozzle, incidenting on a cathode at an inclined angle. The open symbols for Zn-Ni plating in Fig. 5 depict general trends for inclined fluid injection. Starting with a low nickel content and a high overall rate constant in the immediate neighbor of the impingement point, Ni content in alloy deposits rises and the overall rate constant decreases until both reach the plateau values for fully developed turbulent flow. This would imply that a down-stream transition of velocity profile from a wall-jet flow to a turbulent pipe flow occurs gradually.

In respect of Zn-Fe alloy electrodeposition, a general trend on predominant alloy phase transformation, namely η phase $\rightarrow$ δ_1 phase $\rightarrow$ Γ_1 phase $\rightarrow \Gamma$ phase with an increase in Fe content of deposits, can be seen as mass transfer rate at the cathode is decreased. Operational current density should be kept over 600 mA/cm^2 to obtain uniform Fe content in deposits as is illustrated in Fig. 4. It is also clearly seen in Fig. 5 that there is a significant difference between wall-jet flow and parallel flow in relation to streamwise uniformity of alloy deposits. For a parallel flow, Fe contents in alloy film did not virtually vary with location along the cathode, except for a slight decrease in Fe composition up to 5 hydraulic diameters downstream scaled from the point of impingement for inclined flow injection. Solid symbols for Zn-Fe plating in Fig. 5 depict general trends for inclined fluid injection. Starting with a low value in the very immediate neighborhood of the impingement point, Fe content in alloy deposits rises and reaches a plateau, regardless of supporting electrolytes examined.

Addition of aluminium sulphate markedly diminished the non-uniform distribution of Fe contents along the length of the cathode, as shown in Fig. 5, which might be attributed to the buffer reaction;

$$Al_2(SO_4)_3 + 2OH^- \rightleftharpoons 2Al(SO_4)OH + SO_4^{2-} \qquad [1]$$

in the vicinity of cathode surface.

The rate of increase in Fe content slackens with increasing downstream distance, and ultimately iron content seems to approach the value for fully developed pipe flow with identical Reynolds number.

The incorporation of trace metallic species in the electrodeposits may be either diffusion controlled or activation controlled(15). Codeposited contents of minor elements in Zn-Fe alloy deposits were linearly dependent on their bulk concentrations in sulphate electrolyte as shown in Fig. 6. Incorporation of minor species, therefore, appears to be a first-order rate process with respect to their bulk concentrations, and thus can be expressed by the following equations:

Flux of trace species j is given by,

$$N_j = \bar{k} C_\infty \qquad [2]$$

where, C_∞ stands for its bulk concentration, for which overall heterogeneous rate constant $\bar{k}$, reaction rate constant k_R and mass-transfer coefficient k_M, may be written

$$\frac{1}{\bar{k}} = \frac{1}{k_R} + \frac{1}{k_M} \qquad [3]$$

$$k_R = k_R^0(C_0) \cdot \exp[-\alpha FE/RT] \qquad [4]$$

Influence of ionic migration on the mass-transfer can be neglected in the presence of excess supporting electrolyte, thus under given plating conditions,

$$k_M = D_j/\delta_c = \phi(Re, Sc) \qquad [5]$$

Since flux of major constituents, Zn and Fe, is much larger than that of minor species, the composition of trace elements is then,

$$N_j/N_{Zn\text{-}Fe} \simeq \bar{k} C_\infty / i_T \lambda \eta \qquad [6]$$

As can be seen from Eqs. (2) - (6), the rate of codeposition is governed by overall rate constant $\bar{k}$, which depends on harmonic mean of kinetic constant k_R and mass-transfer coefficient k_M. Hence, it is likely that low $\bar{k}$ values for Mn, Ni, In and Sb indicate slow electrochemical kinetics well below diffusion limiting current. Whereas noticeably high $\bar{k}$ value for Tl is seen probably owing to its high diffisivity, and a slight shift towards higher value of $\bar{k}$ for Cu may be caused by additional cementation effect(16). Dimensionless correlation equation for fully developed turbulent flow in a rectangular flow-channel based on Chilton-Colburn analogy can be written as,

$$k_M = 0.023(D_j/d_e)Re^{0.8} Sc^{1/3} \qquad [7]$$

therefore, mass transfer limiting current is given by,

$$i_L = ZFC_\infty k_M \qquad [8]$$

Figure 7 provides dependence of partial current density of species associated with alloy plating on total current density. Limiting current density for Zn^{2+} and H^+ was calculated respectively from Eqs. 7 and 8. Substitution of values; $Re_{avg} = 2.1 \times 10^4$, $d_e = 2.1$ cm, $\nu = 0.026$ cm^2/s, $D_{Zn^{2+}} = 1.4 \times 10^{-5}$ cm^2/s, $D_{H^+} = 1.8 \times 10^{-4}$ cm^2/s, $C_{Zn^{2+}} = 7.65 \times 10^{-4}$ mol/cm^3, $C_{H^+} = 4.0 \times 10^{-5}$ mol/cm^3,
gives,
$i_{L,Zn^{2+}} \simeq 1050$ mA/cm^2, $i_{L,H^+} \simeq 115$ mA/cm^2.
Note that calculated diffusion limiting current density for H_2 evolution is far below observed values, which can be reasonably interpreted in terms of acceleration of mass-transfer by intensive gas bubble agitation at the cathode surface.

It is generally known that metal electrodeposits tend to grow in dendritic and/or powdery form, at limiting current density(17), which result in so-called non-adhesive "burnt" coatings. Neither of electroplated specimens in Fig. 7 exhibited "burnt" appearance even at $i_T = 1600$ mA/cm^2, where measured $i_{Zn^{2+}} = 898$ mA/cm^2 approaches fairly close to the calculated $i_{L,Zn^{2+}} = 1050$ mA/cm^2.

Consequently, it is likely that actual limiting current for Zn^{2+} and H^+ would be enormously higher, presumably by a factor of $2 \sim 5$, than the previously calculated values.

CONCLUSION

Counter-flow injection at the lowest angle possible towards a strip surface within the appropriate range of Reynolds number could facilitate both high and uniform mass-transfer in a cell, which would lead to uniform alloy plating quality. Alloy plating is greatly affected by generation of hydrogen bubbles on the cathode surface as a simultaneous side reaction at high current densities of practical interest. Although gas evolution may seem to be disadvantageous from a current efficiency standpoint, a significant enhancement of mass transfer was observed through the measurements of incorporation rates of trace indicator ions in alloy deposits.

NOMENCLATURE

C_∞ bulk concentration (g/cm^3, mol/cm^3)
C_o electrode surface concentration (g/cm^3, mol/cm^3)
d_e hydrodynamic equivalent diameter (cm)

D_j diffusion coefficient of species j (cm^2/s)
E electrode potential (V)
E^0 standard electrode potential (V)
F Faraday constant 96484.6 (C/equiv.)
i_T total current density (A/cm^2)
i_j partial current density (A/cm^2)
i_L limiting current density (A/cm^2)
$i_{L,j}$ limiting current density of species j (A/cm^2)
k overall heterogeneous rate constant (cm/s)
k_M mass-transfer coefficient (cm/s)
k_R electrochemical kinetic constant (cm/s)
k_R^0 standard rate constant ; k_R for $E = 0$, (cm/s)
N_j flux of species j $(g/cm^2 \cdot s,\ mol/cm^2 \cdot s)$
R gas constant 8.31441 $(J/mol°K)$
T absolute temperature $(°K)$
X coordinate in flow direction and distance from the incident point (cm)
X_w weight fraction of Fe in Zn-Fe alloy deposit
Z charge number (-)
α transfer coefficient (-)
δ_c Thickness of concentration boundary layer (cm)
ε gas bubble enhancement factor (-)
η current efficiency (-)
λ a proportional constant $(g/A \cdot s)$
ν kinetic viscosity (cm^2/s)
Re Reynolds number $(U\ d_e/\nu)$
Re_{avg} average Reynolds number $(\bar{U}\ d_e/\nu)$
Sc Schmidt number (ν/Dj)
Sh Sherwood number $(k_M\ de/Dj)$

REFERENCES

(1) R. Noumi, H. Nagasaki, Y. Hoboh and A. Shibuya: SAE Tech. Pap. Ser. 820332 (1982).

(2) T. Adaniya, T. Hara, H. Fukushima and K. Higashi: Tetsu-to-Hagané, **69** (1983), p.959.

(3) T. Tsuda, K. Asano and A. Shibuya: Trans. ISIJ, **26** (1986), p.53.

(4) T. Tsuda, A. Shibuya, M. Nishihara, K. Yamada, M. Katoh and K. Yanagi: Tetsu-to-Hagané, **72** (1986) 8, p.58.

(5) B. Meuthen, J. H. Meyer zu Bexten and D. Wolfhard: AESF 5th Continuous Strip Plating Symposium, (1987).

(6) A. Kodama, A. Matsuda, T. Yoshihara and H. Kimura: AES 4th Continuous Strip Plating Symposium, (1984).

(7) Raggio, Ulvieri, Raso and Ramundo: AESF 5th Continuous Strip Plating Symposium, (1987).

(8) T. Tsuda, A. Shibuya, M. Nishihara, F. Terasaki, K. Yamada, M. Katoh and K. Yanagi: Proc. AES 71st Annual Technical Conference, American Electroplaters Society, Inc., New York (1984) C7.

(9) A. Weymeersch, R. Winand and L. Renard: Plating and Sur. Fin., **68** (1981), p.56.

(10) H. M. Wang, S. F. Chen, T. J. O'keefe, M. Degrez and R. Winand: J. Appl. Electrochem., **19** (1989), p.174.

(11) P. V. Shaw, L. P. Reiss and T. J. Hanratty: AIChEJ, **9** (1963), p.362.
(12) E. M. Sparrow and N. Cur: J. Heat Transfer, **104** (1982), p.82.
(13) T. Sydberger and U. Lotz: J. Electrochem. Soc., **129** (1982), p.276.
(14) R. N. Sharma and S. V. Patankar: Int'l J. Heat Mass Transfer, **25** (1982), p.1709.
(15) L. F. G. Williams: J. Electrochem. Soc., **126** (1979), p.566.
(16) J. D. Miller: Miner. Sci. Eng., **5** (1973), p.242.
(17) N. Masuko and T. Tsuda: Future Perspectives in Electrochemical Processing and Technology, in "New Trends and Approaches in Electrochemical Technology", Chapter 1, Kodansha-Scientific (1993).

Table 1.   Total Quality Management from R & D phase through Mass-Production phase

[Phase-I] : Laboratory scale experiments
- Basic developments on products & process towards built-in quality assurance by underlying engineering principles

[Phase-II] : Bench scale &/or pilot plant experiments
- Product performance evaluation through trial production
- Verification and correction of detailed process & hardware design

[Phase-III] : Construction & Start-up
- Process optimization, reliability and reproducibility
- Tune-up & modification of hardware and control systems
- Learning curve for operation & maintenance skills

[Phase-IV] : Commercial Production
- Creative team activities by all
- Utilization of quality control methods for process improvements (i.e. Know-'how' statistically, not always necessary to know-'why' technically)
- Integrated accumulation of small individually, yet important, improvements in process, quality and delivery.

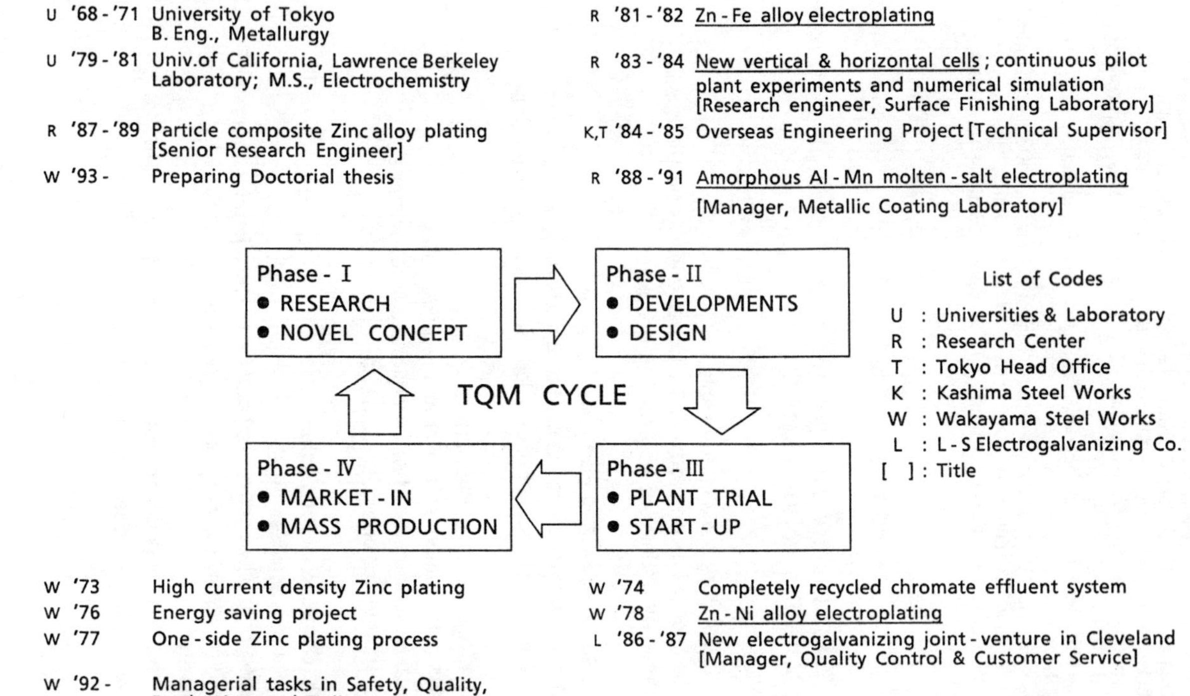

Table 2.   TQM PHASE CYCLE and the SPEAKER'S CAREER PATH

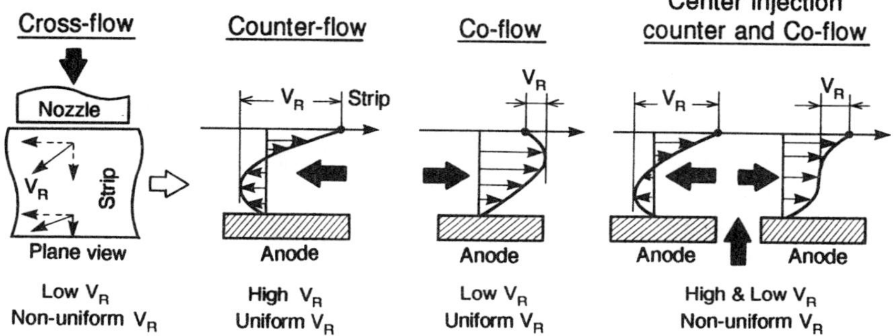

Fig. 1. Electrolyte injection models of various relative fluid velocities represented in a coordinate system with a moving strip

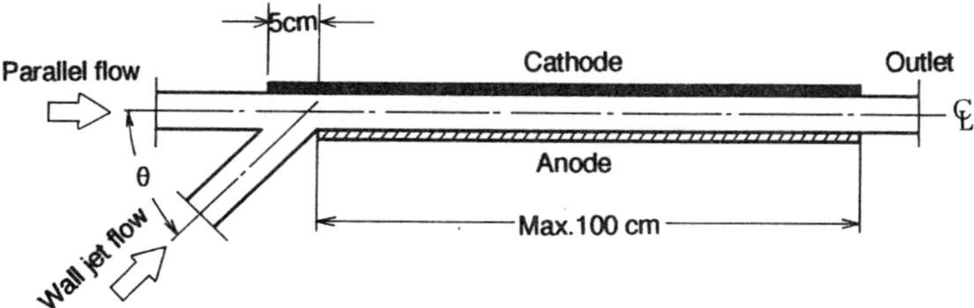

Fig. 2. Schematic view of rectangular flow channel cell with variable incident angle

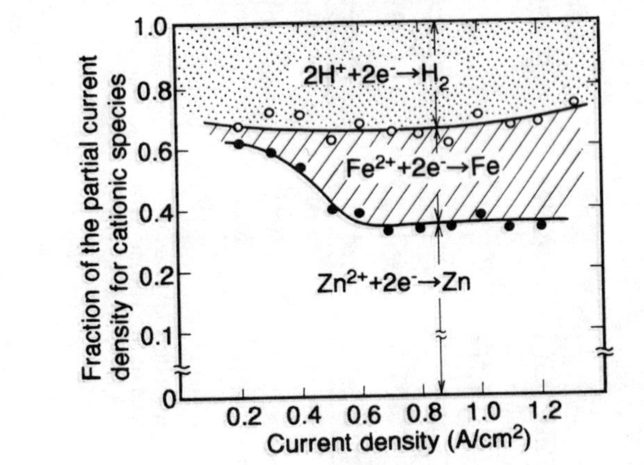

Fig. 3. Influence of Reynolds number on Ni content in deposit

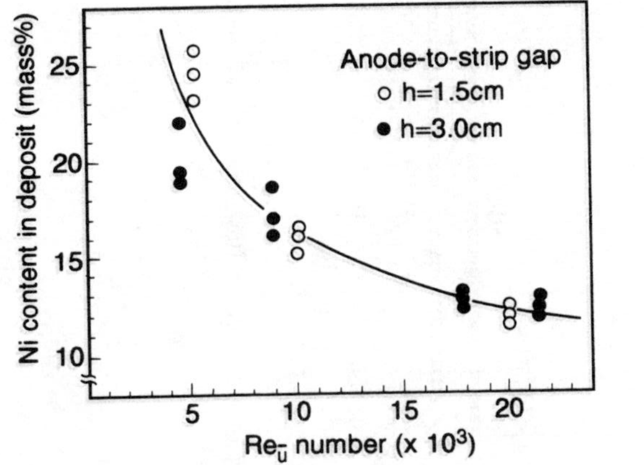

Fig. 4. Effect of current density on alloy content and current efficiency of Zn-Fe electrodeposition

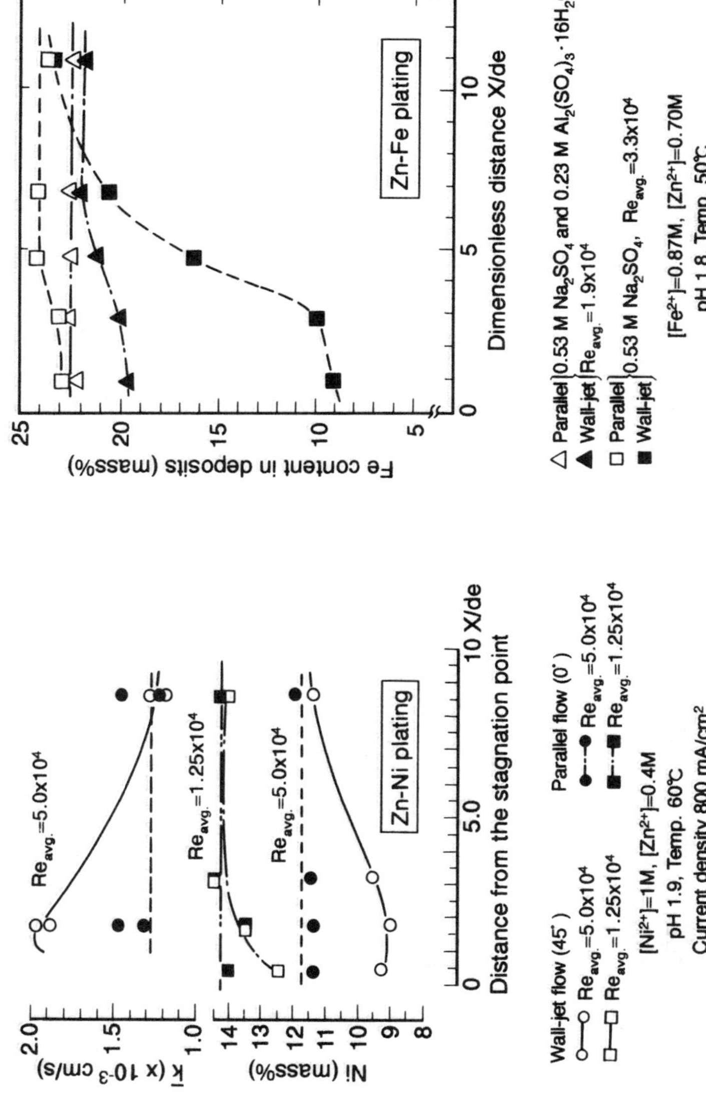

Fig. 5. Comparison of streamwise deviation in alloy content between wall-jet flow and parallel flow systems

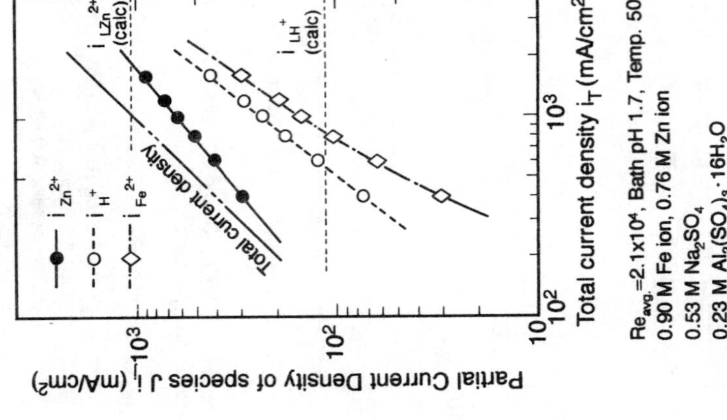

Fig. 7. Variation of partial current densities with total galvanostatic current

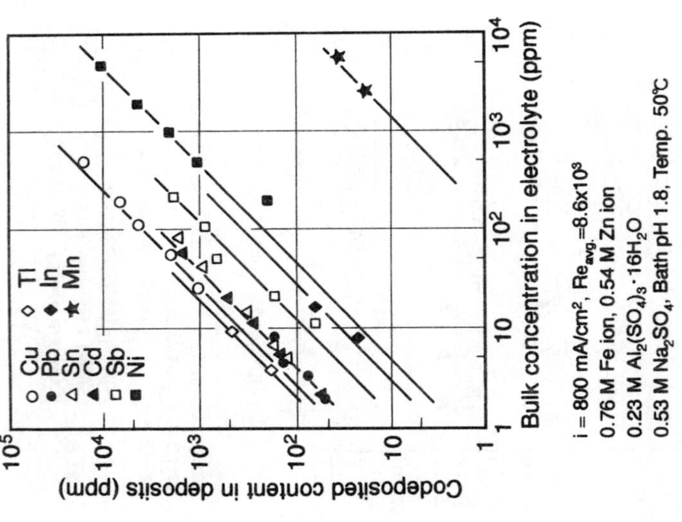

Fig. 6. Effect of minor species concentration on codeposited content in Zn-Fe layer

IN-LINE THIN ORGANIC FILM COATING FACILITY IN THE ELECTROGALVANIZING LINE

Kazuya Ono
Kawasaki Steel Corporation Chiba Works
1, Kawasaki-Cho, Chuoh-ku, Chiba 260, JAPAN

A new in-line coating facility was installed in the electrogalvanizing line of Kawasaki Steel Chiba Works in August 1988.
This line is incorporated several automatic controlling features in (1) coating temperature and viscosity, (2) coating weight, (3) strip temperature, (4) on-line fluorescent X-ray coating amount gage, (5) on-line infrared ray film thickness gage.
This revamping work made possible an in-line treatment of PLASCOAT K4, with a record monthly production of 8,000T in half a year following the start-up of the in-line operation in October 1988. The current monthly production amounts to 15,000T.

1. INTRODUCTION

In August 1988, the in-line coating facility in the electrogalvanizing line of Kawasaki Steel Chiba Works was extensively modified to manufacture PLASCOAT K4 coated steel sheets for automobiles by replacing the old simple coating facility with a full-scale composite line capable of both electrogalvanizing and coating treatments. In order to manufacture PLASCOAT K4[1] in a single line, it was necessary to substantially revamp the old coating facility.
PLASCOAT K4, which has been developed as a coated steel sheet for automobile use, is a filmy organic composite-coated steel sheet that is given Zn-Ni alloy plating, a chromate coating, and finally an organic- resin coating containing colloidal silica. PLASCOAT K4 has outstanding corrosion resistance , although its coating thickness is only $1\,\mu$ m. It also provides high spot weld-ability.[1]
Within 6 months after commencing the in-line produdion of PLASCOAT K4, the monthly production was 8,000T, which at present reaches 15,000T. This report describes the in-line coating facility and its operating conditions.

2. EXAMINATION OF THE OBJECTIVES

2.1 Quality Requirements for PLASCOAT K4

PLASCOAT K4, which has been developed as a coated steel sheet for automobile use, is a filmy organic composite-coated steel sheet that is given Zn-Ni alloy plating, a chromate coating, and finally an organic-resin coating containing colloidal silica. The cross-sectional
structure is shown in Fig. 1. PLASCOAT K4 has outstanding corrosion resistance, although its coating thickness is only $1\,\mu$ m. It also provides high spot weld-ability. Further, in order to facilitate its application to the bake hardenability (BH) steel sheet, it is designed so that resin baking can be achieved

by short-term heating at 150℃ or less.

2.2 Technical Problems with Filmy Coating

In the past, the electrogalvanizing line had a roll coater for finger print resistant steel sheet use, but in order to meet the high-level quality required for PLASCOAT K4, the following new techniques needed to be developed:
(1) High-accuracy coating thickness control
(2) High-accuracy strip temperature control
(3)On-line coating thickness measurement
In particular, in the continuous line for plating and coating, simultaneous control of coating thickness and strip temperature with line speed fluctuations is important, and the facility must provide stable operation without lowering the productivity of the electrogalvanizing line.

3. OUTLINE OF THE FACILITY

An outline of the new electrogalvanizing line is shown in Fig. 2, the right side showing the existing electrogalvanizing line and the left side , the newly added coating line. The electrogalvanizing line is based on the CAROSEL system of U.S. Steel (currently USX) to which our own special improvements were applied. (2) This electrogalvanizing line was completed in 1982 and commenced operation with four plating cells. Later, the number of cells was increased to seven, before the coating line was revamped in August 1988 to complete the present in- line electrogalvanizing and coating line. Table 1 shows the basic specifications of the line, and Fig. 3 shows an outline of the in-line coating facility. Fig. 4 shows the manufacturing process for PLASCOAT K4, which is the principal product from this line.

3.1 Paint Supply Facility

The paint supply facility consists of the paint warehouse, paint mixing tank and paint circulation tank. To save labor automatic management of the paint warehouse and automatic control of paint viscosity are carried out, and an outline of the automatic paint bath control system is shown in Fig. 5. In order to achieve high-accuracy control of the coating weight, controlling the paint viscosity is indispensable, while paint temperature control, which greatly affects viscosity, is also important. The present system provides a heating system which uses warm water, for both the paint mixing tank and paint circulating tank, and automatically controls the temperature of the paint bath to a constant level. The paint circulating tank is also provided with an on-line viscometer and, while continuously measuring the paint viscosity, a diluting solution is automatically supplied to control the viscosity at a constant level.

3.2 Reverse Roll Coaters

The roll coaters consist of No.1 coater for chromate processing and No.2 coater for resin coating, each coater having two heads on the top surface side and two heads on the bottom surface side, thereby providing eight coaters in total. Fig. 6 shows a schematic diagram of the roll coaters. The two-roll reverse coating method consisting of the applicator roll (APR) and pickup roll (PUR) is employed, but the system is also designed to perform natural coating.

3.2.1 Roll speed control

The speed of APR and PUR is controlled to a high accuracy by an AC motor fitted with variable voltage/variable frequency (VVVF) control.

3.2.2 Inter-roll pressure control

The inter-roll pressure of the coater greatly affects the coating weight, and is particularly important for adjusting the paint coating-weight profile in the strip width direction. In the present facility, a load cell is provided between each respective set of two rolls (between PUR and APR, and between APR and BUR) to detect the inter-roll pressure, and a servo-motor is installed to maintain the inter-roll pressure at an appropriate level.

3.2.3 Coating weight control

The line uses the two coating weight control formulas shown in Table 2. Formula No.1 is a linear model, while formula No.2 is non- linear, and the use of these formulas enables the coating weight to be controlled to a constant value against fluctuations in line speed by changing the speed of either APR or PUR. In Fig. 7, a comparison is shown between the calculated and measured values of the resin coating weight when formula No.2 was used, and illustrates the high control accuracy of the coating weight.

3.3 Oven and Cooler Facilities

The oven and cooler facilities comprise No.1 oven and No.1 cooler for chromate use, and No.2 oven and No.2 cooler for resin coating. Table 3 shows the specifications of these facilities.

3.3.1 Ovens

Circulated hot-air gas jet heating is used, and line speed fluctuations are compensated by varying the hot air jet speed with the opening of the circulated hot-air damper. Fig. 8 shows the logic for controlling the temperature of the sheet. The strip temperature is controlled by setting up the necessary hot air flow rate from such data as the oven gas temperature, line speed, and strip thickness, and adjusting the opening of the circulated hot air damper.

3.3.2 Coolers

No.1 cooler for the chromate treatment is aircooled, while No.2 cooler for resin coating uses watercooling. The strip temperature in No.1 cooler is controlled in the same way as that in the ovens by the damper opening, after setting up the required cooling air flow rate from such data as the cooling air temperature, line speed, and sheet thickness. No.2 cooler provides rapid cooling by a high-pressure water spray.

3.4 Computer Control System

An outline of the computer control system used in this line is shown in Fig. 9. When the new coating facility was installed, electrical DDC (direct digital control) and instrumentation DDC were added. Electrical DDC is applied for line operation control, catenary control, coater roll speed control, inter-roll pressure control and paint coating weight control in the coating section, and instrumentation DDC is applied for oven control, cooler control and strip temperature control. Each control system operates through a CRT display to save labor and rationalize the operation of the line.

3.5 On-line Quality Assurance Equipment

To ensure quality along the entire length of the product coil, on- line analysis apparatus is provided. The line employs on-line flourescent X-ray analysis to measure the alloy-plated coating weight and Cr coating weight, and on-line infrared (IR) analysis to measure the resin coating weight.

3.5.1 On-line fluorescent X-ray analyzing system

The line uses the same on-line flourescent X-ray analyzing system to measure the alloy-plated coating weight that has been used since earlier days.[3] An outline of this system is shown in Fig. 10. This system provides continuous measurement of the alloy-plated coating weight and Cr coating weight, thereby giving quality monitoring along the entire length of the product coil.

3.5.2 On-line IR analyzing system

On-line measurement of the resin coating weight is done by an IR analyzing system; an outline of this system is shown in Fig. 11. Its operating principle lies in measuring the absorbance peak of infrared rays caused by the C-H stretching vibration in the resin coating film, and converting the degree of absorbance into the resin coating weight. This system uses a special filter for enhancing the measuring accuracy.

4. OPERATING CONDITIONS

4.1 Production

PLASCOAT K4 from this line reached a monthly production of 8,000 T as early as April 1989, 6 month after starting operation. The present monthly production is 15,000 T or more, showing that it has been possible to apply mass-production techniques within a short time.

<u>4.2 Quality</u>

<u>4.2.1 Coating weight</u>

Fig. 12 shows the frequency distribution of the Cr and resin coating weights. The Cr coating weight is within ± 20 mg/m^2 of the target value, and the resin coating weight is within ± 0.1 g/m^2, illustrating the high level of coating weight control. Fig. 13 shows the fluctuation in Cr and resin coating weights when the line speed is changed. When the line speed is changed from 120 m/min to 80 m/min, and then from 80 m/min to 45 m/min, the Cr coating weight fluctuation is within ± 10 mg/m^2 and resin coating weight fluctuation is within ± 0.05 g/m^2. This illustrates that coating weight control, which has been a technical problem, is now satisfactory.

<u>4.2.2 Baking temperature</u>

The frequency distribution of baking temperatures of products is shown in Fig. 14. Both No.1 and No.2 ovens show satisfactory strip temperature control to within ± 10℃ of the target value. Fig. 15 shows the fluctuation of No.2 oven outlet temperature after line speed changes. The response speed for sheet temperature control is satisfactory, resulting in a temperature within ± 5℃ of the target value.

5. CONCLUSIONS

This report provides an outline and operating conditions of the in-line thin coating system which has been added to the electrogalvanizing line at Chiba Works.
(1) PLASCOAT K4, which has been in production on this line since October 1988, recorded a monthly production of 8,000 T in 6 months, and at present is achieving a monthly production of 15,000 T, illustrating the establishment of mass-production techniques within a short period.
(2) The quality of PLASCOAT K4 is also satisfactory, and it is possible to control the Cr weight coating to within ± 20 mg/m^2 of the target value, the resin coating weight to within ± 0.1 g/m^2, and the baking temperature to within ± 10 ℃.
(3) The coating weight control and strip temperature control, which have given technical problems in the past, are now satisfactory, and it has become possible to control the Cr coating weight to within ± 10 mg/m^2 of the target value, the resin coating weight to within ± 0.05 g/m^2, and the paint baking temperature to within ± 5 ℃ of the target value.

6. REFERENCES

<u>6.1 Journals</u>

1. K. Yamato, T. Ichida, and T. Irie: Kawasaki Steel Giho, 21(1989)3, 216
2. A. Komada, Y. Hirooka, K. Miyaji, T. Yoshihara, A. Matsuda, H. Yasunaga, and H. Kimura: Kawasaki Steel Giho,15(1983)1,1
3. K. Yamamoto, W. Tanimoto, N. Makiishi, Y. Matsumura, Y. Makino, and H. Abe: Kawasaki Steel Giho, 21(1981)2,107

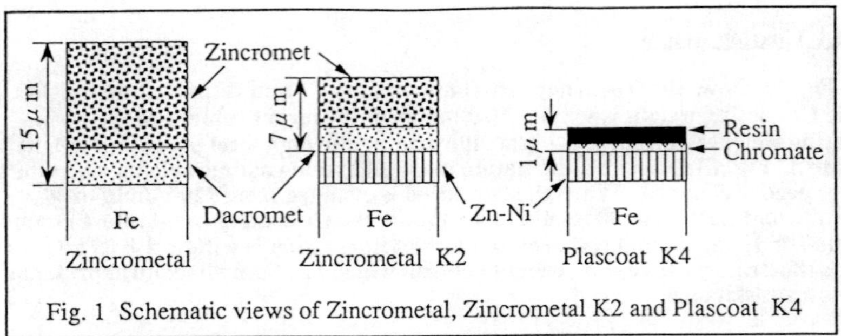

Fig. 1 Schematic views of Zincrometal, Zincrometal K2 and Plascoat K4

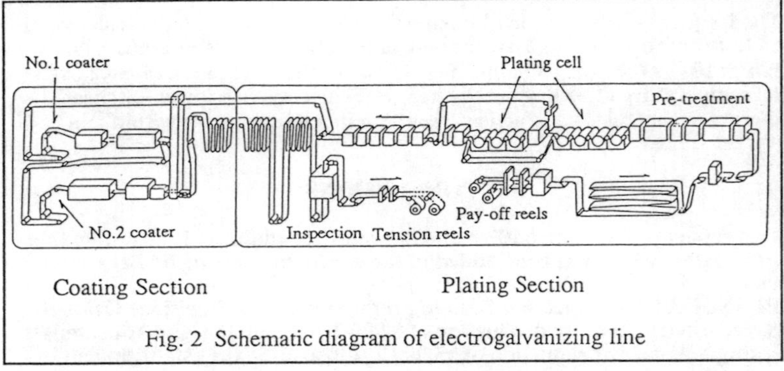

Fig. 2 Schematic diagram of electrogalvanizing line

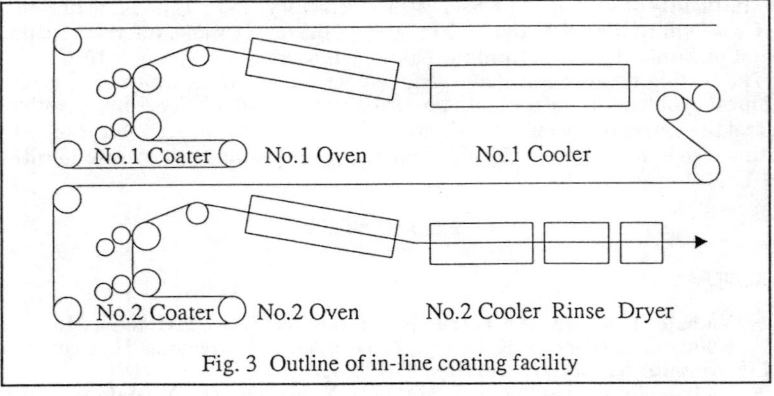

Fig. 3 Outline of in-line coating facility

Table 1 Line Specifications

Item	Specification
Production capacity	25,000 t/month
Line speed	
Entry	150 m/min (max)
Center	120 m/min (max)
Delivery	150 m/min (max)
Sheet thickness	0.4 ~ 1.6 mm
Sheet width	760 ~ 1,830 mm
Coil weight	
Entry	42 t (max)
Delivery	25 t (max)
Plating cell	CAROSEL type 7 cell (25 KA×20 V×14 units)
Coating equipment	
Coater	2 roll reverse coater × 8 units
Oven	Circulated hot air jet
Cooler	No.1:Air cooler
	No.2:water cooler

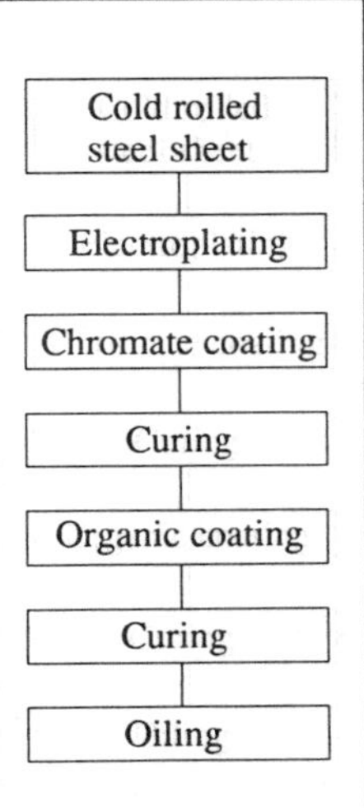

Fig. 4 Manufacturing process

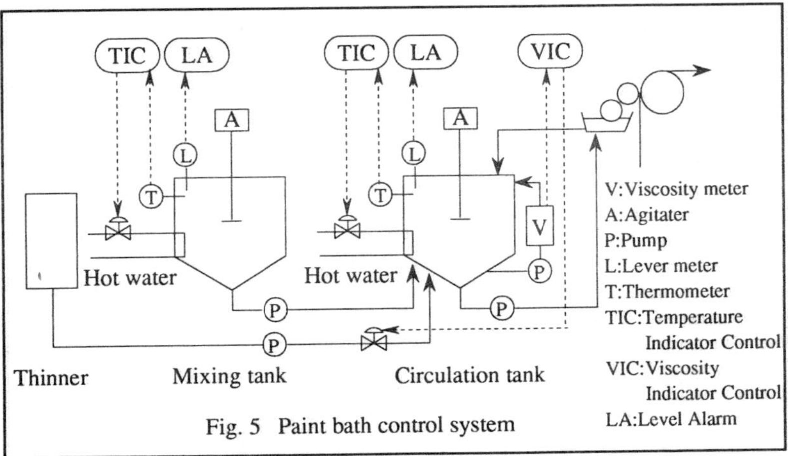

Fig. 5 Paint bath control system

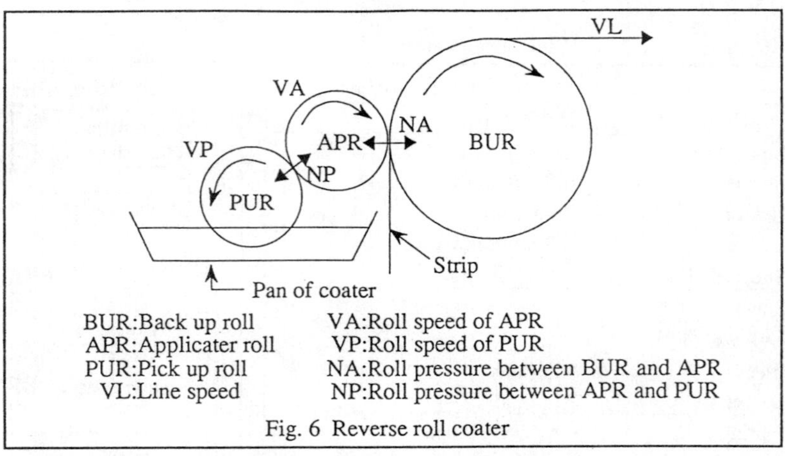

Fig. 6 Reverse roll coater

Table 2 Coating weight control formula

Model	Formula	Note
Model formula No.1 (Linear model)	$L = A*VA+B*VP+C*VL+D$	L:Coating weight VA:Roll speed of APR VP:Roll speed of PUR VL:Line speed A,B,C,D,K,x,y,z:Coefficients
Model formula No.2 (Non-linear model)	$L = K * VA^x * VP^y * VL^z$	

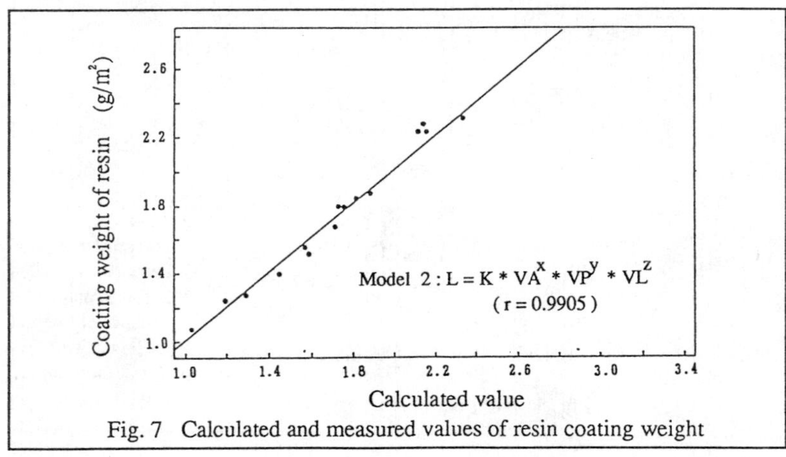

Fig. 7 Calculated and measured values of resin coating weight

Table 3 Oven and cooler specifications

	No.1 oven	No.1 cooler	No.2 oven	No.2 cooler
Heating method (cooling)	Circulated hot air jet	Air jet	Circulated hot air jet	Water spray
Strip conveying	Catenary	Floater	Catenary	Catenary
Furnace length	7.5 m×2 zone	10 m×3 zone	7.5 m×3 zone	5 m
Temperature	Max 500℃	15～30 ℃	Max 500℃	20 ℃

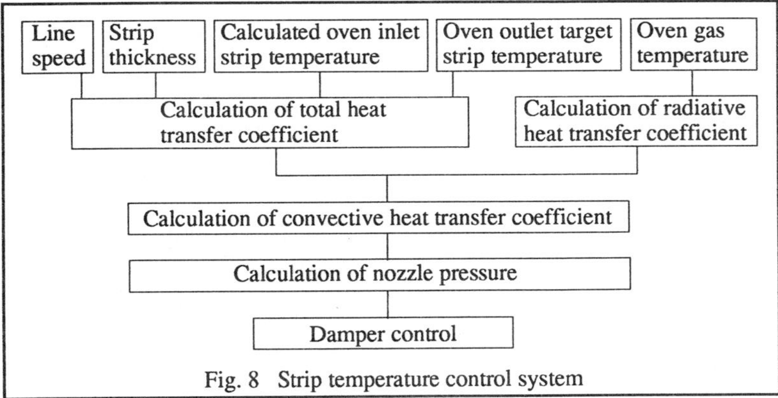

Fig. 8 Strip temperature control system

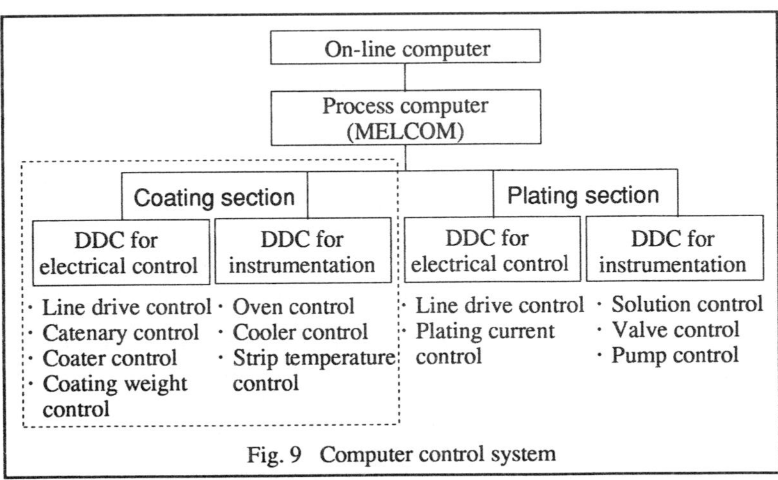

Fig. 9 Computer control system

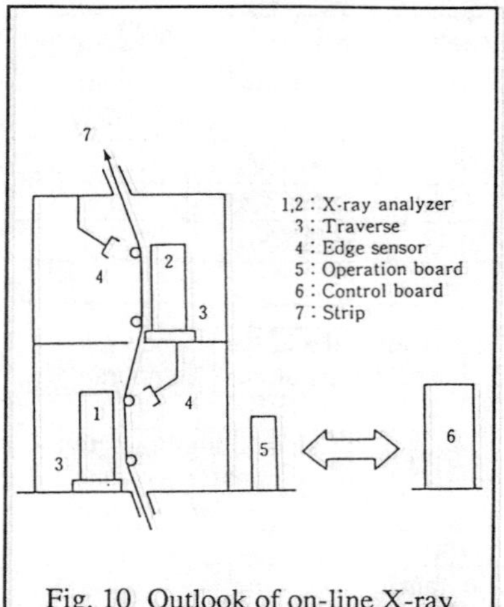

Fig. 10 Outlook of on-line X-ray analyzing system

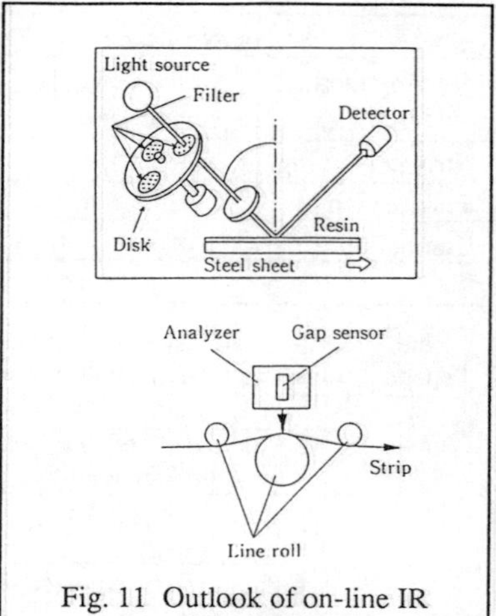

Fig. 11 Outlook of on-line IR analyzing system

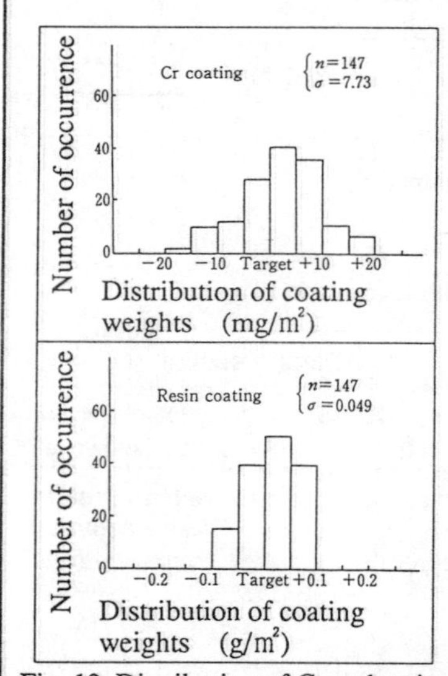

Fig. 12 Distribution of Cr and resin coating weights

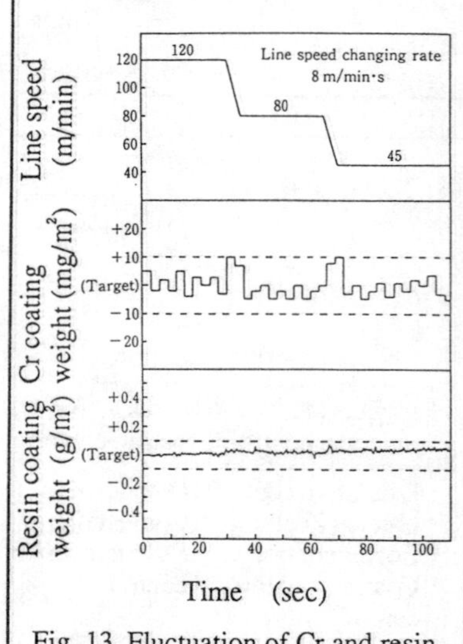

Fig. 13 Fluctuation of Cr and resin coating weights when line speed changes

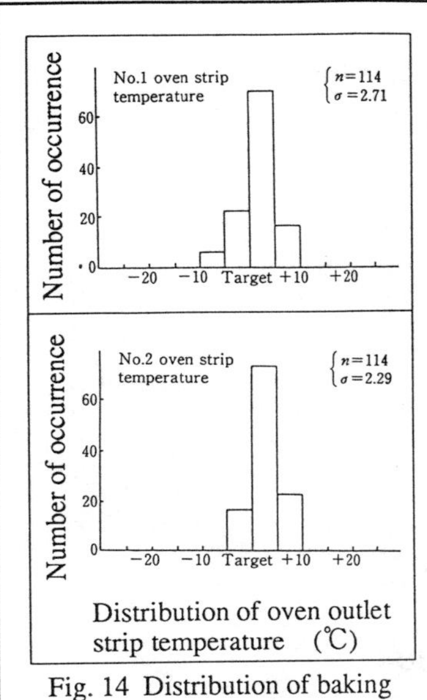

Fig. 14 Distribution of baking temperature

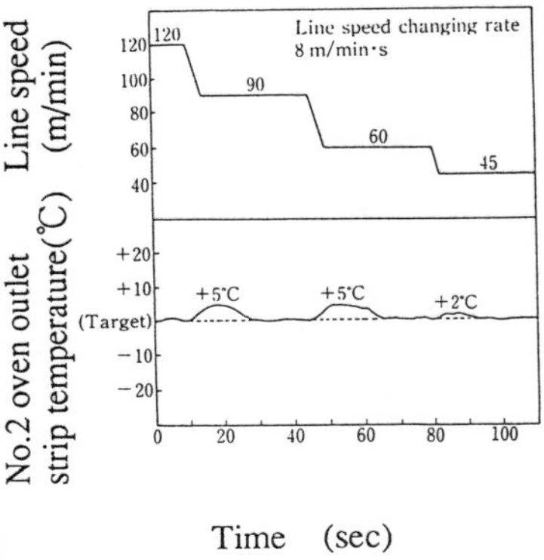

Fig. 15 Fluctuation of No.2 oven
outlet strip temperature
when line speed changes

GLUE ANALYSIS IN COPPER ELECTROREFINING

Carol A Davis and Greg A Hope
Griffith University
Nathan, Qld 4111
Australia

The measurement of glue levels in acid sulphate copper electrorefining electrolyte has been achieved with an accuracy of better than $\pm$ 0.2ppm using a portable glue monitor. Half lives for the loss of glue due to acid hydrolysis at a range of temperatures have been determined, yielding an Arrhenius activation energy of approximately 130 kJmole^{-1}. The effect of cathode adsorption of glue and selected water soluble polymers during the electroplating process on the crystal structure of the copper deposit has been investigated by scanning angle X-ray diffraction. Highly oriented copper deposits have been generated from electrolytes containing polyvinylalcohol and polyethyleneimine indicating that crystal face specific adsorption must occur for these polymers. Copper deposits from electrolytes containing glue are not as highly oriented, but there is still evidence for specific adsorption of the additive.

INTRODUCTION

In the acid sulphate electro-refining process used to further purify smelted copper, organic reagents are added to prevent nodulation of the cathode deposits, to control the physical and chemical properties of the refined copper and to help control the insoluble slimes that form (1-6). These additives are present in very low levels, approximately gram per tonne level and small variations in the quantity of the additives has serious consequences for plant economy and tankhouse operation. The significant effect that the two major additives, currently in use in most refinery tankhouses, gelatin glue and thiourea, (leveller and grain size modifier, respectively) have on the quality and crystal structure of the resulting copper deposit is reasonably well documented, however their modes of action are not well understood.

Thiourea is employed for its properties of grain refinement and deposit brightening. Knuutila et al report that thiourea enhances the initial nucleation of the deposit at the cathode surface (5). They claim that the specific adsorption via the sulphur onto specific crystal faces refines the grain size by obstructing the deposit growth in certain directions. In acid sulphate solutions, thiourea forms stable complexes with univalent copper species. However ambiguity exists over the exact co-ordination number of the ligand to the cuprous ion.

Gelatin (glue), is the hydrolysis product of collagen and is an amorphous composition of protein fragments ranging in size from simple dipeptides to large multichain polypeptides. Each collagen macromolecule contains three polypeptide chains, twisted about one another. Gentle heating causes these asymmetric, helical macromolecules to denature to random-coil gelatin chains. Gelatin is further hydrolysed in acid solution whereby the peptide chains (-NHCO- bonds) are partially destroyed and terminal groups such as RNH_3^+ are produced.

Glue (gelatin) assists in the production of a level, dense copper deposit with improved purity (1). It is believed that the glue acts by specific adsorption onto active growth sites on the cathode, temporarily blocking deposit growth (2). Some authors suggest that it is the ionic character of the molecule in acidic environments that is responsible for this adsorption (3). During the electro-refining process the glue is consumed by acid hydrolysis, incorporation into the deposit and by adsorption onto anode slimes (4). It is therefore necessary to continually dose the system with glue to maintain optimum plating conditions. As various factors affect the consumption of the reagent, constant addition rates do not necessarily ensure a constant or optimum cell concentration of glue. Additive levels are generally monitored in the refinery by inspection of the physical appearance of the cathode copper. A further complication is that the composition of gelatin glue, due to the nature of its source, may vary from batch to batch . It has a finite shelf life and is also susceptible to bacterial attack. Degradation from bacterial attack can generate products that have an adverse effect upon the plating process.

The refining process has developed empirically in the industrial environment and a result of this has been the creation of a viable commercial process with restricted understanding of the mechanism by which the additives operate in the plant solutions. The electrochemistry has also been complicated by the concentrated solutions (approximately 180 gl^{-1} H$_2$SO$_4$, 45-50 gl^{-1} Cu^{2+}) used in the refinery. Therefore, great emphasis has been placed upon developing new techniques that will, under tankhouse conditions, enable the monitoring of single additives directly and easily.

A number of techniques for measurement of the active additive concentration in the tankhouse electrolytes have been proposed (2,3,7-12). Most of these techniques electrochemically measure the change in the cell or electrode polarization with additive identity and concentration, however not all of these techniques can be used in undiluted tankhouse electrolyte.

The mechanisms by which additives work can be classified depending upon whether they are chemical or physical plating modifiers. Physical modifiers adsorb onto the surface either specifically or non-specifically and block growth sites. Chemical modifiers alter the electrochemical pathway of the original process by either altering the stability of one or more species within the original route by complexation or by creating a new option in the pathway. These two aspects of plating action are interactive to some extent, either due to adsorption of the additive complex onto the cathode surface or the presence of some complexing functional groups on the predominantly non-complexing adsorbent. Glue would be classified as a physical modifier as it adsorbs onto the cathode surface (the adsorption strength is field dependent). The strongest adsorption would be expected to occur at those surface sites which are likely to grow the fastest, thus effectively blocking the electron transfer at dendrite growth centres. Thiourea exhibits elements of both classifications. It modifies the electron transfer mechanism and also adsorbs on the cathode surface affecting the crystal grain size.

It is well known that the adsorption of organic additives at a cathode surface has the capacity to increase the cell overpotential (1). A portable Glue Monitor that is based upon cell overpotential measurements has been constructed. It has been sucessfully plant tested and gives a direct measure of the glue concentration in the tankhouse electrolyte.

EXPERIMENTAL

The role of glue in the electrorefining process has been examined using scanning angle X-ray diffraction (XRD) techniques and the glue monitor and related overpotential measurement techniques. XRD was employed to investigate the crystal orientation of the copper deposit generated from electrolyte containing glue and without glue. As the loss of glue in the plating process is predominantly through acid hydrolysis, the glue monitor was employed to measure the half lives of active glue in tankhouse electrolyte at a range of temperatures that encompass those normally experienced in the refinery. An Arrhenius plot was constructed to calculate the activation energy for this hydrolysis pathway.

The information reported here is specifically oriented towards copper electro-refining, however the methods and techniques developed here (eg Glue Monitor and related overpotential measurement techniques) should be valuable for any commercial metal electroplating process (such as Cr, Co, Ni, Zn etc).

Reagents

The tankhouse electrolyte and glue (Davis Gelatine, technical grade AM1, average molecular weight 60k) were kindly supplied by the Copper Refineries Pty Ltd, Townsville. The electrolyte supplied varied between 45-50 gl^{-1} Cu^{2+} and 185-195 gl^{-1} H_2SO_4. The electrolyte was 'de-glued' by boiling for 30 minutes to hydrolyse any residual glue (3). The residual thiourea in the electrolyte appeared to be unaffected by this treatment. The glue monitor response (GMR) was checked to verify full hydrolysis of the active glue and generate the base GMR.

Glue Monitor Technique

The nature and effect of solution flow on the copper electrode are important parameters which affect the quality and rate of copper plating. Initial work indicated that the cell overpotential measured at a stationary cathode was dependent upon the solution flow patterns or stirring rate. Employing a rotating disk as a cathode, sets up steady hydrodynamic flow across the electrode face allowing additive concentration influences on the cell overpotential to be studied under uniform hydrodynamic conditions. A fully operational, plant tested, portable glue monitor based on this technique has been constructed. The concentration of glue in the tankhouse electrolyte can be determined (to within 0.2 ppm) and monitored, using this prototype. Development of an on-line system is being undertaken.

A typical run involved the equilibration of the acrylic cell containing tankhouse electrolyte in a waterbath, at constant temperature, followed by the determination of the steady state GMR, suspended in the cell electrolyte. Fresh electrolyte was required for each successive trial. If no other contaminants were added to the system, the electrolyte was recycled (no more than twice) using the de-gluing process.

Calibration of the glue monitor was achieved using the method of standard additions. A stock glue solution of appropriate concentration was made up and stabilized by the addition of a small amount of copper sulphate. Known amounts of stock glue solution were added to a de-glued electrolyte solution. The steady state GMR was plotted versus the cell glue concentration to generate the calibration curves. The effect of temperature on the calibration curve was assessed by completing this exercise at a range of temperatures. Half-lives were determined at a range of temperatures by monitoring the GMR over time for a solution initially dosed with a known aliquot of stock glue solution. From this information, the dose rates required to achieve particular active glue levels for the constant current plating experiments conducted at various temperatures were determined.

<u>Constant Current Plating</u>

The ultimate test of additive interaction is the structural and chemical quality of the plated copper. Constant current test plating enables this to be evaluated directly. Copper deposits (approximately 5 x 2cm) were plated from electrolytes of various compositions. One plate was generated from deglued electrolyte that had been dosed with glue to maintain a level of 2ppm, and another from deglued electrolyte. Two further deposits were produced from deglued tankhouse electrolyte, dosed with polyvinylalcohol (10ppm 15k PVA) or polyethylenimine (1ppm 40k PEI). The samples were plated at 65°C for approximately six hours at 250 Am^{-2} (except for the PVA dosed experiment which was plated at 350 Am^{-2}). The deposits were washed in dilute acid and distilled water before XRD investigation.

A Phillips PW 1820 X-ray generator, operating at 40kV and 40mA was used for the XRD investigations. A cobalt radiation tube (wavelength 1.790Å) was used as the source. The reflected radiation was monitored with a counter connected to a computer under the control of a Sietronics SIE 112 software package. The 2-theta scan range was from 40° to 100°. The scan rate was 2° per minute with a step size of 0.05°.

RESULTS

An overpotential versus time curve is generated by the Glue Monitor. Figure 1, a typical GMR curve, clearly demonstrates that glue adsorbs onto the cathode surface. There is an initial drop in overpotential whilst the nucleation on the surface occurs, followed by an increase in potential as the surface coverage by the glue increases over the growing nuclei. The steady state potential reached is proportional to the amount of surface blockage achieved by the concentration of active glue present.

Standard additions of glue, to deglued electrolyte, generates a calibration curve relating the steady state Glue Monitor Response (GMR) to the concentration of glue. Figure 2 indicates that the GMR is relatively linear up to 2ppm for copper electroplating at 65° C. Calibration runs were completed at temperatures ranging from 55° to 70^oC which encompass most tankhouse operating temperatures. The electrolyte temperature has a large effect on the range and linearity of the GMR (Figure 3). The optimum operating range for the glue monitor indicated by this data requires temperatures within a few degrees of 65^oC and glue concentrations below 2ppm. These ranges are comparable to most refinery systems.

The Glue Monitor development has facilitated studies on aspects of the acid hydrolysis of active glue.The acid sulphate refinery electrolyte will cause hydrolysis of the gelatin chains to continue and an increase in the rate would be expected with an increase in cell temperature.The hydrolysis of gelatin in the tankhouse electrolyte will continually decrease the molecular weight of the macromolecular chains. A point will be reached when the molecular weight drops below a threshold value required to generate a measurable effect upon the overpotential of the system and no longer have any levelling effect upon the deposit topography. The 'active' concentration of glue in the system is then defined as the concentration of species that have molecular weights greater than the threshold value. The GMR is related to the active concentration of glue in the system. The kinetics of the hydrolysis of glue in the tankhouse electrolyte, as determined by the Glue Monitor investigations, will represent only the active concentrations of glue present. These calculations will generate information specific to the refinery operations, as the active concentrations are what they are concerned with. The 'real' kinetic information, however will be closely related as much of the information is determined at reasonably short periods of time with respect to the determined half lives and therefore the active and real glue concentrations will not be greatly different.

Half lives of 5.2 hours at 55°, 4.4 hours at 60° , 2.7 hours at 65° and 0.63 hours at 70°, were determined in copper refining electrolyte for active glue (Figure 4). Assuming a first order hydrolysis rate, an Arrhenius plot generated an activation energy of approximately 130 kJ $mole^{-1}$ (Figure 5). This compares with a value of 83 $kJmole^{-1}$ that was previously determined for the acid hydrolysis of animal glue (collagen) (13). Saban et al determined this value by high performance size exclusion chromatography (13). They used synthetic electrolyte (150 gl^{-1} H_2SO_4 and 46 gl^{-1} Cu^{2+} as $CuSO_4$) that was of a different composition to the present study, lower in both acid and copper concentration. The hydrolysis rates that they determined will reflect the lower acid concentration. As operating temperatures in the copper refining plant can vary between about 63° and 68°, quite different hydrolysis rates will exist across the range of temperatures in the cells, thus monitoring of cell glue concentration should be an important part of refinery quality control procedures.

Additive adsorption onto the cathode surface has been shown to generate specific topographical effects by scanning angle X-ray diffraction (Figure 6). Copper deposits from acid sulphate electrolyte, de-glued and containing glue were subjected to XRD investigations. Electrolyte containing glue (and residual thiourea) generates a copper deposit that produces a mixture of the 111, 220 and 200 crystal plane peaks in the XRD spectrum. The 220 peak is enhanced compared with randomly oriented copper powder. However, the 111 and 220 crystal planes of the 'de-glued' electrolyte deposit have grown relatively faster in preference to the almost non-existent 200 plane (Figure 6). This would imply that there is some specificity in the adsorption of the glue moiety on to the cathode surface. More dramatic control of the crystal face orientation of the copper deposited occurs with low levels of some water soluble polymers (water soluble polymer additives are widely used in copper plating for the manufacture of printed circuit boards (14,15)) such as polyvinyl alcohol (PVA) and polyethyleneimine (PEI) (Figure 6). The PVA can be seen to significantly inhibit the growth of the 111 and the 200 crystal faces, as the diffraction from the 220 crystal faces dominate the XRD spectrum. PEI adsorption generates another pattern of crystal face orientation. Growth of all faces except the 111 are inhibited. An element of specific adsorption by these additives is evident in this behaviour. In each of these solutions, the residual thiourea in the tankhouse electrolyte will be competitive in preferential adsorption that may occur onto the cathode surface. Interactions between the additives also may play a role in influencing the competitive equilibria that must exist for the complex cathode adsorption phenomena occurring in this system.

With the development of the Glue Monitor, investigation of other phenomena that affect the cell overpotential, such as substitute additives, is now possible. Ideally a surface active additive that is stable towards acid hydrolysis, has a known standard composition and does not precipitate or adsorb onto slimes would be preferred. The possibility of being able to increase the current density of the plating process by using this substitute additive would also be commericially attractive. To be considered as a viable substitute additive this compound would also need to be cost effective. High molecular weight, water soluble polymers that are acid stable (such as PVA, PEI) have been targeted for trialling as new or substitute additives. The basic families of these polymers (varying molecular weights and degree of hydrolysis) and synthetically modified family branches are being investigated.

Routine monitoring of the tankhouse electrolyte generates data banks for future reference and will enable assessment of cathode quality and current efficiency with respect to active glue levels. The benefits of the glue monitor for quality management are many. This prototype is small in size and totally portable. It can be used anywhere in the tankhouse or laboratory. This technology may be adapted to an online system. This technology affords quick determinations of active glue levels, approximately 20 minutes (initial calibration takes approximately 1.5 hrs but is not required for every determination of glue level). Monitoring of glue stock quality, activity and batch reproducibility is also possible. Estimates of storage life and dose tank lifetimes could be made. The loss of glue in the electrolyte through mechanisms other than acid hydrolysis can also be investigated. For example, in a non-flowing cell, the glue loss in an isolated cell electrolyte sample has been compared with the glue loss due to acid hydrolysis combined with the bulk cell glue loss over time.

ACKNOWLEDGEMENTS

The assistance with and the financial support of various aspects of this research by Copper Refineries Pty. Ltd., the Australian Research Council and Griffith University Research Grants Committee are gratefully acknowledged.

REFERENCES

1. F.A. Lowenheim, Modern Electroplating, p. 32, Wiley Interscience, New York (1974).
2. G. Troch, M. Degrez, and R. Winnand, in *Electrochemistry in Minerals and Metal Processing/1984*, P.E. Richardson, S. Srinivasan, and R. Woods, Editors, PV 84-10, pp. 642-54, The Electrochemical Society Softbound Proceedings Series, Pennington, NJ (1984).
3. T.N. Anderson, R.D. Budd, and R.W. Strachan, Metal Trans. 7B, 333 (1976).
4. S.E. Afifi, A.A. Elsayed, and A.E. Elsherief, J. Metals, 38-41 (1987).
5. K. Knuutila, D. Foresen, and A. Pehkonen, *The Effect of Additives on the Electrocrystallisation of Copper*, J.E. Hofmann et al, Editors, p 142, Metallurgical Society, Warrendale, PA (1987).
6. R. Winand, G. Toch, M. Degrez, and P. Harlet, Industrial Electrodeposition of Copper. Problems Connected to the Behaviour of Organic Additions, in *Proceedings of the International Conference on the Application of Polarization Measurements in the Control of Metal Deposition*, ASM, Victoria B.C., Canada (1982).
7. M.L. Rothstein, W.M. Peterson, and H. Siegerman, Plat. and Surf. Finish., 78 (1981).
8. N.R. Bharucha, Z. Zavorsky, and R.L. Leroy, Met. Trans. 9B, 509 (1978).
9. H.S. Jennings and F.E. Rizzo, Met Trans. 4, 921 (1973).
10. R. Haak, C. Ogden, and D. Tench, Plat. and Surf. Finish., 52 (1981).
11. L.F. Toth, PhD Dissertation, Univ. of Missouri-Rolla (1975).
12. S. Krzewska, L. Pajdowski, H. Podsiadly, and J. Podsiadly, Met. Trans. 15B, 451 (1984).
13. M.D. Saban, J.D. Scott, and R.M. Cassidy, Met. Trans. 23B, 125 (1992).
14. J.P. Healy, D. Pletcher, and M. Goodenough, J. Electroanal. Chem. 338, 155 (1992).
15. M. Yokoi, S. Konishi, and T. Hayashi, Denki Kagaku 52, 218 (1984).

Figure 1
Glue Monitor Response Curve

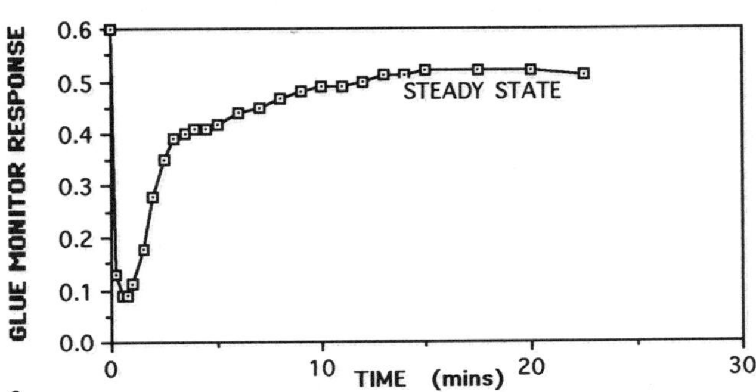

Figure 2
Glue Monitor Calibration Curve at 65ºC

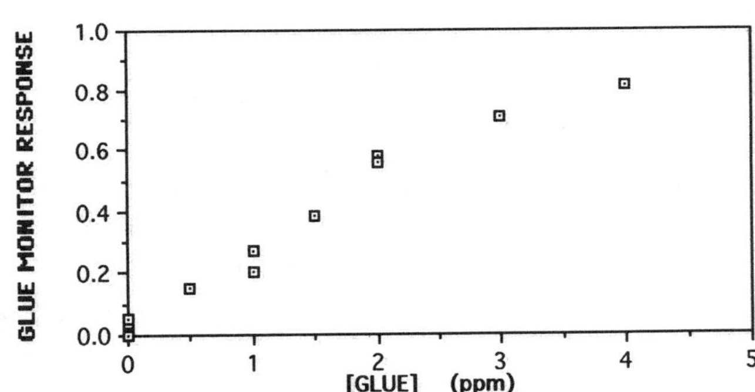

Figure 3
Glue Monitor Calibration Curves at Various Temperatures (55-70ºC)

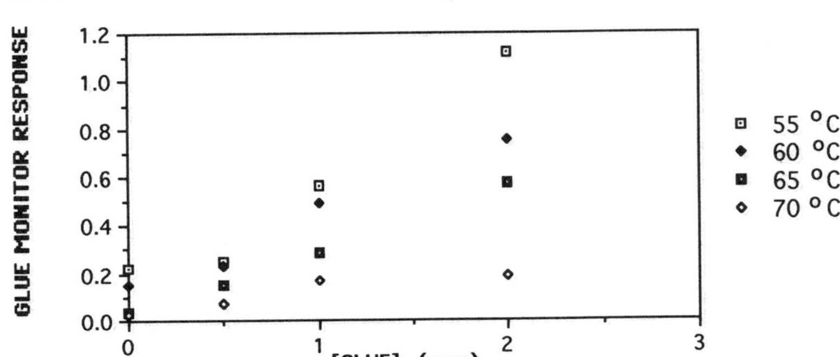

Figure 4
Half-Lives of Active Glue in Copper Electrorefining Acid Sulphate
Electrolyte at a Range of Temperatures

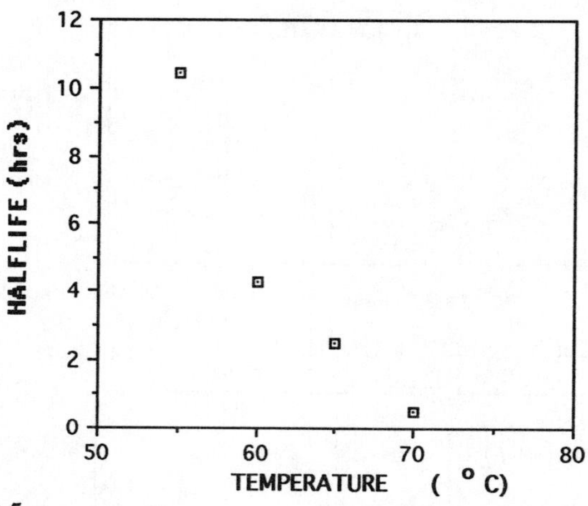

Figure 5
Arrhenius Plot for Hydrolysis of Active Glue in Copper Electrorefining
Acid Sulphate Electrolyte

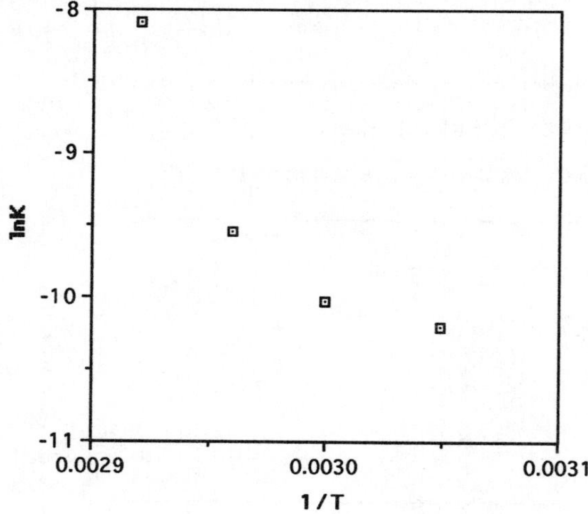

Figure 6
XRD Patterns for Deposits Generated from Electrolyte that is
a) Glue Dosed b) De-glued c) PEI dosed and d) PVA dosed

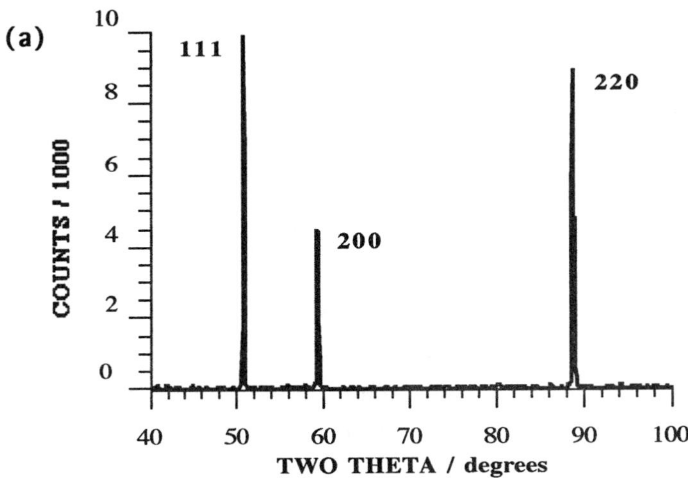

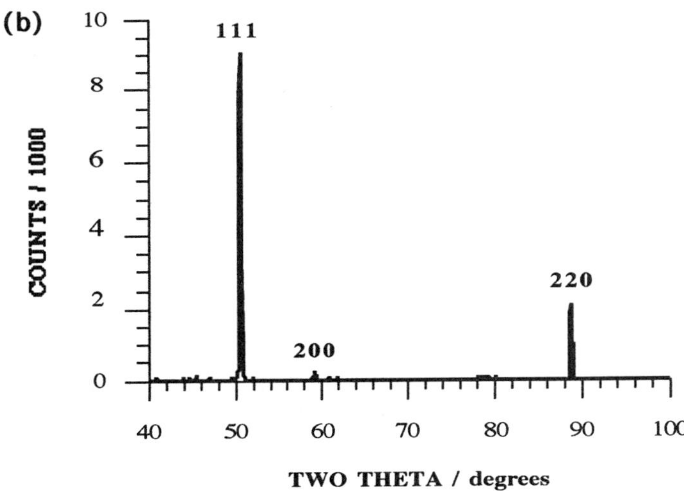

Figure 6, Continued

(c)

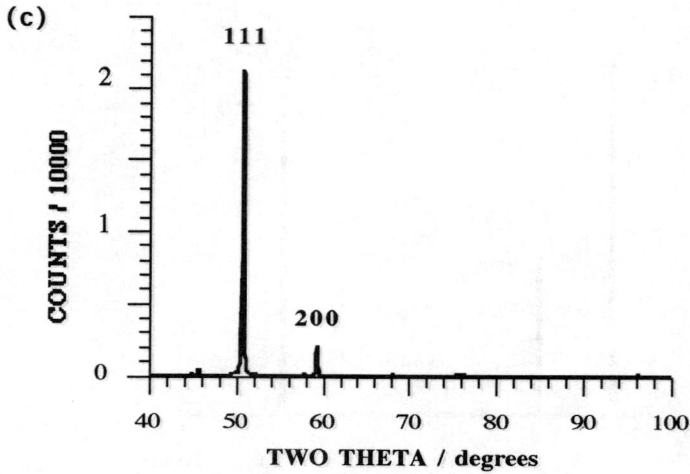

(d)

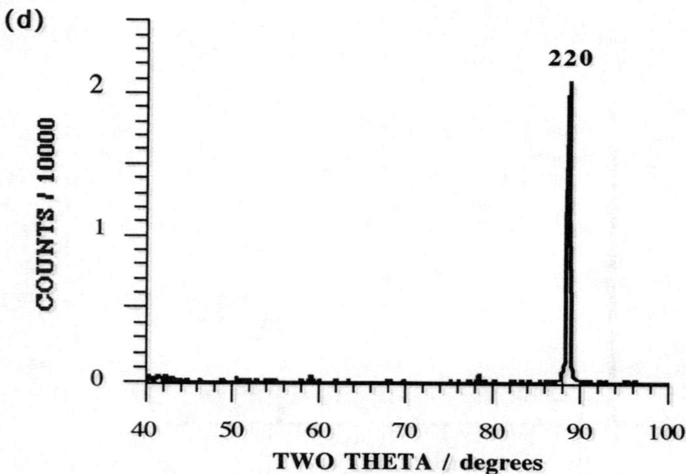

ADDITIVE CONTROL AND COPPER PLATING QUALITY

Gregory A Hope* and Carol A Davis
Science and Technology
Griffith University
Nathan, Qld 4111
Australia

The electrolytic production of copper is a batch process and the optimum concentration of additives should be determined by consideration of the electrolyte composition. Analysis of additives in the electrolyte is a major quality control requirement. Thiourea is an example of an additive which interacts with the electrolyte components. It reacts with cupric solution ions to form complex cuprous cations. The sulfate salt of the complex cation is more soluble than the chloride salt, and elevated chloride levels remove thiourea from the electrolyte. The thiourea species are electroactive and can be formed on a rotating disc electrode. Electrochemical oxidation of the complex results in the formation of a cathode contaminating film which impedes copper deposition. Copper production is also subject to occasional quality control problems which are unrelated to additive levels. The inadvertent introduction of metal carbides into the electrolyte, and the effect of power interruption are considered as examples of stochastic problems encountered in tankhouse operation.

INTRODUCTION

The quality of electrolytically produced copper in the plant is judged on both the physical and chemical properties of the product. The commercial quality is determined by the chemical composition, and copper is one of the highest purity metals traded in commerce. The electrorefining of copper commences with anodes which are generally better than 99.5% pure and aims to produce cathodes with a copper purity of greater than 99.99%. The major requirement is to generate a very pure metal which is suitable for electrical and wire drawing applications. The specification for cathode purity has maximum acceptable levels for problem elements such as bismuth and sulfur. Electrorefining is a batch process with recirculating electrolyte, which leads to the concentrating of soluble impurities from the anodes. There are many aspects to quality control in the refinery, anode uniformity, electrode spacing, electrolysis current profile, nature of the cathode, starter sheet geometry, slime handling, electrolyte flow pattern etc. In this paper we will consider some of the electrochemical aspects relevant to the maintenance of quality in the copper tankhouse.

The electrorefinery electrolyte requires monitoring on a routine basis to ensure that the chemical composition remains within acceptable limits. In addition to the fundamental reagents, such as the free acid and the copper ion concentration , refineries usually find that their electrolyte requires monitoring for particular impurities which are detrimental to the copper refining process. Typically these are elements which are characteristic of the source ores (or scrap), for example arsenic and bismuth. Therefore, quality control in the refinery also has to address the purity and purification processes which operate in the maintenance of the electrolyte composition.

The electrorefining and electrowinning processes utilize additives to improve the quality of the plated copper. The additives work predominantly by controlling the physical structure of the plated copper cathode. There are two basic types of electrochemical additives used in the plating of metals. They are given various names, such as smoothing agents, grain refiners, brighteners and throwing power enhancers but a more general classification describes them as either physical or chemical modifiers, dependent upon their mode of action. Glue and gums are typical physical modifiers which adsorb at the cathode. These macromolecules act as a cathode polarizer, and can assist in the suppression of some undesirable reactions such as hydrogen evolution. Their major role in copper plating is to assist in the generation of a smooth deposit. In acid solutions, the colloid hydrolyses and the molecular weight progressively decreases. This change in the molecular mass reduces the colloid size and is accompanied by a decrease in the extent of the cathode polarization. The hydrolysis may also lead to a roughening of the metal deposit. Analysis of gelatine in the electrolyte has been a continuing challenge to the industry, and we have reported the development (1) of a simple portable Glue Monitor based upon the cell overpotential dependence. The Monitor response is similar to the polarisation data reported by Troch, Degrez and Winand in 1984 (2). In acid sulfate copper refining, the glue used is a technical gelatine with an average molecular weight near 60,000 (3). The quality of the glue is somewhat dependent on the sourcing of the material, and how it is handled. As a natural product, it has a limited shelf life, and bacterial attack has been found to generate products which are detrimental to copper refining. The electrowinning process cannot use glue for smoothing, due to the scumming of the protein associated with solvent carry over. Alternative reagents such as vegetable polysaccharides (eg guar gum) are used when required by electrowinning plants. The practical use of physical modifiers has remained largely unchanged for almost a century in the copper electrorefining industry, where the most significant improvements have been the introduction of dosing techniques, and (more recently) analysis. This is not true of other areas in the copper plating industry. Synthetic water soluble polymers which are widely used in copper plating, particularly printed circuit board manufacture (4,5), appear to be unsuitable or unacceptable to the refining and electrowinning industries.

The major chemical modifiers used in the electrolytic production of copper are thiourea and chloride. These reagents complex with cuprous ions to form species which are relatively stable in the tankhouse electrolytes. The charged complexes formed appear to be concentrated near electrodes where there is a higher level of cuprous ions generated in the electrochemical processes associated with the reduction and oxidation between cupric ions and the copper metal. Thiourea is effective in modifying the the crystal structure of the cathode copper. It is utilised in both electrowinning and electrorefining operations as an effective grain refining additive and is generally added at a rate which maintains the solution content at less than 5 ppm as free thiourea. Overpotential measurements indicate that the chemical modifiers do not significantly impede the commercial cathode reduction process, and chloride ions may assist the anode corrosion. The chemical modifiers can also complex with other metals and reagents. These reactions can be beneficial or detrimental to copper quality depending upon the nature of the process. The interaction of reagents is an important aspect of quality control, which has not been extensively studied.

EXPERIMENTAL

The electrolyte and materials used in the investigations described in this paper have been obtained from the tankhouses where the research was initiated. Some investigations have been repeated in several tankhouses. Commercial electrolyte has been used whenever possible during laboratory investigations. The composition of the commercial electrolyte is dependent upon the refinery and day to day conditions, but in general the composition is kept near 200 gpL sulfuric acid and between 40 and 50 gpL of copper. In some experiments it was necessary to use a model electrolyte in order to minimise the interaction of the electrolyte components and interference during electrochemical measurements. The synthetic electrolyte used in these experiments was 45 gpL Cu^{2+} as the sulphate, and 180 gpL H_2SO_4 prepared from Ajax analytical grade reagents in de-ionized water. Glue was obtained from the refineries stock of commercial grade AM1 technical gelatine supplied by Davis Gelatine. Thiourea was technical grade reagent supplied by Ajax Chemicals. VC, NbC, and MoO_3 were obtained from Aldrich Chemicals. A chromium carbide (Cr_3C_7) was synthesized by melting chromium and cast iron in an argon arc and dissolving the iron matrix in dilute hot sulphuric acid to leave the carbide as an insoluble residue. Rotating ring disc electrode (RRDE) measurements utilised a Pine instruments AFRDE4/AFMSRX system. Cyclic voltammograms were obtained using a PAR 273A Potentiostat/Galvanostat. The glue analysis instrument was constructed in-house. All electrochemical measurements were undertaken on electrolyte which had been heat treated for 30 minutes at a temperature greater than 90 °C, to hydrolyse residual colloidal gelatine material(6). The X-ray Photoelectron Spectroscopy (XPS) spectra were recorded using a Perkin -Elmer 560 ESCA / SIMS double-pass CMA with an angle resolvable aperture; Mg Kα X-ray radiation was obtained from a dual Mg anode source operated at 500 W, 15 kV. The band pass energy was 100 eV for survey and 50 eV when recording individual peaks. The photoelectron signal obtained from a freshly sputtered gold coupon was used to calibrate the spectrometer in the region of interest. The Au 4f $_{7/2}$ peak was assumed to occur at an energy 83.86 eV (7). Measurements were made under control of 560 MACS version VI software, and data was transferred to Macintosh Classic II for processing.

RESULTS

The control of copper quality requires the maintenance of the tankhouse on a routine basis with respect to the major components and additives in the electrolyte, together with the less common investigation and rectification of perturbations in the plating system. Thiourea is considered as an example of the routine additive control and several other examples of plating quality problems are presented.

<u>Thiourea</u>

Thiourea (Tu) reacts in acid sulfate copper plating electrolyte to form various complex species. There are several cuprous thiourea complexes formed, and some of the electrochemical behaviour of thiourea in the presence of copper ions has been reported. In figure 1 we present the Xray crystal structure of a compound formed by the reaction

185

of thiourea with cupric ions in 2M sulfuric acid (8). The crude product collected from the reaction was recrystallised from hot (70 ºC) 2M sulfuric acid. It can be seen from this crystal structure that thiourea coordinates to a copper atom through the sulfur. Four copper species are located in a tetrahedral arrangement at the centre of seven thiourea ligands. The charge is balanced with two sulfate anions, which are not coordinated to the copper thiourea complex. It therefore has four positive charges on the copper thiourea complex. There are two types of thiourea ligand in this structure, six of the ligands are coordinated through the sulfur to two copper atoms, and one which is coordinated to only one copper atom. There is one water of crystallisation in the unit cell of this compound. It would appear that the copper species are either three or four coordinate to the sulfur atoms. The complex is moderately soluble in hot 2M sulfuric acid. The same compound was isolated from solutions with a ten fold excess of cupric ions, and from thiourea solutions into which a small quantity of cupric ion was added. Given this structural information, it is likely that in the tankhouse, the thiourea which is effective in modifying the copper plating structure exists coordinated in a large complex copper (I) cation in the electrolyte. The positively charged cuprous thiourea complex would tend to concentrate near the cathode as the ion transport process proceeds. The formation of formamidine disulfide may occur during the reduction of the cupric species to cuprous ions, as postulated by Pajdowski and others (9,10), but this species is uncharged and may not complex to copper ions.

The valency of the copper species in the compound was investigated by inspecting the freshly ground crystalline compound by (XPS). The spectrum obtained was consistent with bulk cuprous thiourea sulfate. This spectrum is presented as trace (A) figure 2, and the presence of only two singlet copper 2p photoelectron peaks at binding energies of -933 eV and -953 eV is further evidence that copper (II) species, which exhibit paramagnetic splitting of the 2p peaks, are not present in the crystals of this complex. This copper (I) complex was air and solution stable and no special precautions were required in the handling of the crystals for study. Upon addition of chloride ions (as 1M sodium chloride) to a sulfuric acid solution of the complex a white compound precipitated. This compound exhibited a very low solubility in aqueous and polar organic solvents, and was unable to be recrystallised. XPS investigation revealed the presence of chloride anions in the precipitate, and from the spectra it was determined that the sulfate anions had been replaced by chloride ions. Inspection of the data showed the presence of the Cl 2p peak at -199.5 eV, this can be observed in the spectrum shown as trace (B) in figure 2. Under higher resolution, the S 2p multiplexed signal exhibited two distinct peaks for the sulfur atoms in the sulfate complex. One at -172 eV (uncorrected) is assigned to the sulfur atoms in the sulfate, and the other S 2p peak at -166.4 eV was derived from the thiourea ligands. These peaks can be observed by inspecting un-massaged data in figure 3, where the multiplexed S 2p XPS spectra from the sulfate (A) and chloride (B) complexes are presented. Raman spectroscopy supported the XPS data and it can be seen from the spectrum presented as figure 4 that no peaks were present which could be assigned to a sulfur - sulfur bond which would have been expected from any formamidine disulfide species present in the precipitate. The C-N peaks are shifted to slightly higher wavenumbers (1400 - 1600 region), a sulfate peak appears near 980 cm^{-1} and the C-S peaks move to lower wavenumbers in accordance with the changes to the bond strengths of the thiourea molecule on coordination to the copper. The Cu-S peak is located at 236 cm^{-1}.

The interaction of the thiourea complex with chloride has implications for the control of thiourea levels in the presence of elevated chloride concentrations. The low solubility of the complex may remove chloride and thiourea from the electrolyte, thus giving operators the impression that thiourea can correct for high chloride levels. In some plants reverse polarity is applied to improve the quality of the cathode structure. During the anodic pulse, the reagents at the cathode are subject to oxidising conditions, which could decompose the thiourea complex. These conditions can be investigated using cyclic voltammetry (CV) at stationary electrodes or more effectively, by using an RRDE system. The CV at a platinum electrode in synthetic tankhouse electrolyte at 1000 rpm and low ring sensitivity exhibits a single anodic peak for the oxidation of copper metal to cupric ions. In the presence of thiourea the copper to cupric peak is reduced in magnitude and up to three poorly defined peaks anodic of this value develop. This behaviour was continued when the electrode was rinsed and placed into clean synthetic electrolyte free of thiourea. This thiourea derived film could be wiped from the electrode surface with paper tissue and the initial CV characteristic of a clean copper sulfate was regenerated.

The disc electrode was swept from -580 mV to +780 mV vs a Hg/Hg_2SO_4 reference electrode (680 mV vs SHE). This covered the regions in which copper is plated onto the cathode disc from the electrolyte and dissolved from the anode into the electrolyte. In the presence of thiourea we are interested in the generation of cuprous ion species, so in these experiments the platinum ring electrode is held at a potential which will oxidise cuprous ions to cupric species (100 mV vs SHE). The ring electrode is therefore acting as a solution detector for cuprous ion species, in this case, those generated during the reactions occurring at the disc electrode. The relatively long lifetime of aqueous cuprous ion species in the proximity of the electrode surface facilitates the operation of this system (11). This is due to the passage of some of the reduction intermediate cuprous ions past the surface under the force of the hydrodynamic flow generated by the rotating electrodes. A typical RRDE graph for synthetic electrolyte is presented as figure 5. The disc trace resembles a normal CV for copper in sulphuric acid. The ring electrode trace exhibits only a positive current, and this occurs whenever an oxidative electrochemical reaction proceeds at the ring. It can be seen in this figure, that cuprous ions are oxidised at the ring electrode when copper is being deposited at the disc, and when the copper deposited on the disc is being stripped.

The effect of adding thiourea to the solution can be observed in the RRDE data presented as Figure 6, where the simple copper CV is complicated by dosing with less than 0.25 ppm of the reagent. In this figure the disc scan has been extended to - 640 mV. The effect of thiourea on the CV only becomes apparent if the cathodic excursion exceeds the shoulder region seen on the disc CV cathodic peak. The cuprous species are associated with adsorbed or reacted thiourea, and they particularly involve the anodic reaction(s), since it is disc which becomes contaminated following oxidation, impeding the deposition of copper metal on the following cathodic sweeps. The electrochemical behaviour of thiourea in copper solution has been rather well investigated. Onstott and Laitinen investigated the nature of the cuprous complex by polarography in 1950 (12), and demonstrated that in very dilute copper (I) solutions ($1Cu : 10^3 Tu$) that the one electron redox process occurred exclusively. This verified the absence of cupric ions. They postulated that four thiourea ligands were associated with each copper (I) ion in solution. The dissociation constant of this thiourea complex was calculated to be 4.1 x 10^{-16} which indicated a very stable species. It may be predicted from the data, that when the copper to thiourea concentration ratio approaches the values encountered in the tankhouse of $1 : 10^{-5}$ that the conditions are likely to be different as their equilibrium expression involves the product of the copper ion activity and the fourth power of the thiourea activity.

The RRDE technique can generate conditions where we may control the concentration of cuprous ions at the electrode. There are several cuprous ion oxidation signals present in Figure 7, and some of these are directly related to copper (I) thiourea complexes. It is therefore probable that a range of copper (I) thiourea complexes are present at the disc electrode, however the peak labelled C is evidence for the role of thiourea in the plating of copper at the cathode, since it occurs while the disc is still cathodic, but is only detected on the return sweep. It is the differing stabilities of the cuprous thiourea compounds to the swept potential at the disc that generate the cuprous ions which are oxidised at the ring electrode. The analysis of thiourea in tankhouse electrolyte may be complicated by the presence of several complexes in solution. The optimum concentration of thiourea in the electrolyte may depend upon the copper ion and the chloride ion concentrations. The most appropriate level could be determined from the shape of the RRDE response, and the amplitude of the peaks.

The interaction of thiourea with a copper cathode surface has been shown to be strongly and uniformly adsorbed onto all copper orientations with no preference for any crystal face. The adsorption interferes with the crystal growth of the copper electrodeposit (13). Inhibition of the nucleation process can be observed in the modification of the copper deposition peak as shown in the disc trace of figures 6 and 7, where a shoulder can be seen on the cathodic peak.

<u>Cathode Structure</u>

Thiourea and glue are both active in influencing the microcrystalline structure of the copper deposit. The copper deposited during normal refining and electrowinning operations is subject to considerable variation in the electrolyte, and frequently significant changes to the additive levels. Copper cathodes are often plated for up to 12 days and in the deposit these variations in the plating conditions can be revealed with appropriate sectioning, polishing and etching treatments. Consider the photograph of the etched copper cathode deposit presented as figure 8. In the photograph we observe the copper microcrystalline structure exhibited by a copper cathode which had plated for 302.66 hours onto both sides of a copper starter sheet. A linear relationship between thickness of the copper plate and time would mean that 1 division on this micrograph would correspond to 24 hours of plating in the tankhouse. The variation in the division width is a reflection of the faster growth of one side of the cathode compared to the other due to the variation in current density. The observed copper crystal structure was consistent with a regular occurrence in the plating process on a daily basis, an observation which was suggestive that a refinery practise was responsible for the variation. Some of these events were more predominant and easily visible than others. Careful inspection of the microcrystalline structure reveals that the abrupt changes were associated with the end of a period of coarsely crystalline plating. The refinery reported that there were no planned periodic events in the system operation, so it was concluded that the behaviour was indicative of a quality control problem in the reagent make-up and/or dosing system. Inspection of the dosing pumps revealed the need for recalibration. Occasionally some cathodes in this refinery were found to passivate for a period, and when copper deposition recommenced, nodular growth would take place. These occurrences appeared to coincide with the plating faults, and frequently the copper cathode would de-laminate in that region. A typical region is labelled **A** in figure 8. The cathode copper was frequently well outside the purity specification in these regions, which could lead to rejection of the entire section of cathode production, when the fault was extensive.

The inspection of the macro-etched copper cross-section in a scanning electron microscope revealed finer faults in the copper structure. This can be observed in the micrograph presented as figure 9. These faults appear as parallel bright lines in the micrograph and were caused by a rectiformer fault which interrupted the power to the cells. Each time the power fails, it requires a re-start procedure in which the current density slowly recovers. This procedure did not appear to significantly alter the growth pattern of the copper crystals in the deposit, but sufficient variation to the crystal structure occurred enabling the selective etching of the copper in that region of the cross section. EDX and Auger analysis of the etched region did not reveal any compositional change in the metal, in contrast to the faults which which were associated with changes to the copper crystal growth patterns.

The occurrence of problems associated with the production of cathode copper can be quite complicated. The cathode production at one refinery was badly affected by an unexpected increase in electrode shorts due to needle-like dendrite formation (3). Refinery staff tracked the cause down to the presence of a hacksaw blade in a cell. This was a somewhat surprising observation. Hacksaw blades are made from tool steel and are about 83% iron and low levels of tungsten, molybdenum, chromium, vanadium and carbon. The iron dissolves in the electrolyte to give ferrous ions, however the electrolyte already contains a substantial quantity of iron. The other metals are present as carbides, which are in the main relatively unreactive. Inspection of the residue remaining from the dissolution of tool steel in the scanning electron microscope showed that the particulate material consisted of a random shapes with particle sizes varying from less than 0.1 μm to several microns. The dendrite promoting material was derived from the particles which were fine enough to be passed through Whatman grade 1 filter paper. Samples were collected on a 0.2 μm micropore membrane filter. This material was predominantly vanadium and chromium carbides with some molybdenum and tungsten present in the residue. Copper plates with poor characteristics could be generated by addition of hacksaw steel, or hacksaw residue, or vanadium or chromium carbide samples to a plating test cell. Soluble compounds of the problem carbide metals did not generate the same behaviour during plating tests. The dendritic plating behaviour could be stopped by filtration through a 0.2 μm micropore membrane, and re-started with the reintroduction of the residue. The effect of the carbides could be observed by following the cell potential under constant applied current conditions. Plating two test cells and adding the residue from the acid dissolution of a hacksaw blade sample to one cell and vanadium carbide to the other cell generated a change in potential upon addition of the solid. The hacksaw residue had a finer particle size compared to the vanadium carbide, and the greatest effect was observed in that cell. The rapid change in cell overpotential was associated with the onset in dendritic growth and reflects a change in the plating behaviour. The rapid potential change associated with the introduction of carbide particles was not observed when the steel hacksaw blade was introduced into a plating cell, probably due to the slow release of particulate material. The effect was shown to be associated with a large range of carbide steels which contained vanadium, chromium and/or molybdenum. Tungsten carbide did not generate dendritic plating behaviour within the concentration range tested.

CONCLUSIONS

The addition of thiourea to tankhouse electrolyte generates stable, soluble cuprous thiourea complexes which are electroactive and modify the plating behaviour of copper from sulfate solutions. Chloride replaces sulfate in the complex and forms an insoluble compound which could be one mechanism by which chloride and thiourea interact to affect the electrodeposition of copper. Reagent addition rates should be adjusted to compensate for changes in the electrolyte composition, temperature and operating conditions. This implies that the industrial plant has the ability to analyse the electrolyte for glue, thiourea and the other components in solution. Quality control can be assisted by regular inspection of cathode copper microcrystalline structure which can reveal problems associated with the plating system. The control systems can deal with the routine, but on occasions problems arise which require special investigation.

ACKNOWLEDGMENTS

The assistance with and financial support of various aspects of this research by Copper Refineries Pty. Ltd., Southern Copper Pty. Ltd., Olympic Dam Operations and The Australian Research Council are gratefully acknowledged.

REFERENCES

1. C.A. Davis and G.A. Hope in *Quality Management in Industrial Electrochemistry /1993*, D. Hall and Y. Kondo, Editors, **PV 93-19, pp 172-182,** The Electrochemical Society Softbound Proceedings Series, Pennington, NJ (1993).
2. G. Troch, M. Degrez, and R. Winand in *Electrochemistry in Mineral and Metal Processing/1984*, P.E. Richardson , S. Srinivasan, R. Woods, Editors, PV 84-10, pp 642-54, The Electrochemical Society Softbound Proceedings Series, Pennington, NJ (1984).
3. J. O'Kane, CRL Townsville, personal communication.
4. J.P. Healy, D. Pletcher, and M. Goodenough, J. Electroanal. Chem. **338**, 155 (1992).
5. M. Yokoi, S. Konishi, and T. Hayashi, Denki Kagaku **52**, 218 (1984).
6. T.N. Anderson, R.D. Budd, and R.W. Strachan, Metal. Trans. B **7B**, 333 (1976).
7. R.J. Bird and P. Swift, J. Electron Spectrosc. Relat. Phenom. **21**, 227 (1980).
8. G.A. Hope and C.L. Raston, unpublished data.
9. A. Szymaszek, J. Biernat, and L. Pajdowski, Electrochimica Acta **22**, 359 (1977).
10. Ph. Javet and H.E. Hintermann, Electrochim. Acta **14**, 527 (1969).
11. G.W. Tindall and S. Bruckenstein, Anal. Chem. **40**, 1402 (1968).
12. E.I. Onstott and H.A. Laitinen, J. Am. Chem. Soc. **72**, 4724 (1950).
13. B. Ke, J.J. Hoekstra, B.C. Sison, Jr., and D. Trivich, J. Elctrochem. Soc. **106**, 382 (1959).

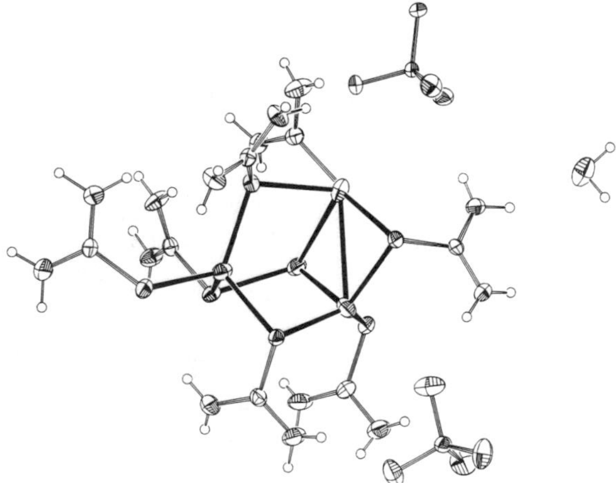

Figure 1. The X-ray single crystal structure determined for Cu$_4$ Tu$_7$ (SO$_4$)$_2$ ·H$_2$O .

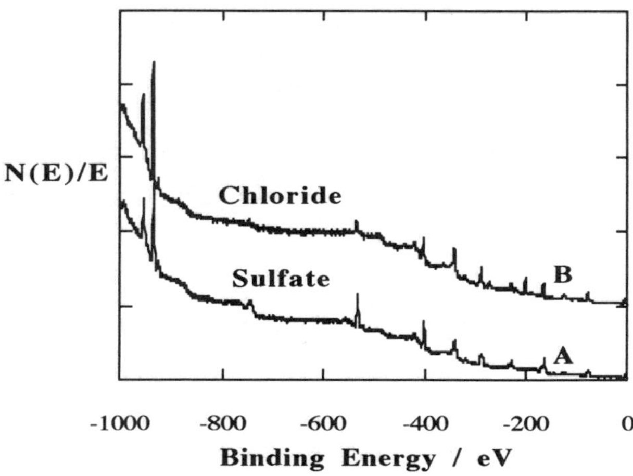

Figure 2. XPS survey spectra of the (A) the sulfate salt, and (B) the chloride salt of the copper thiourea complex isolated from the electrolyte.

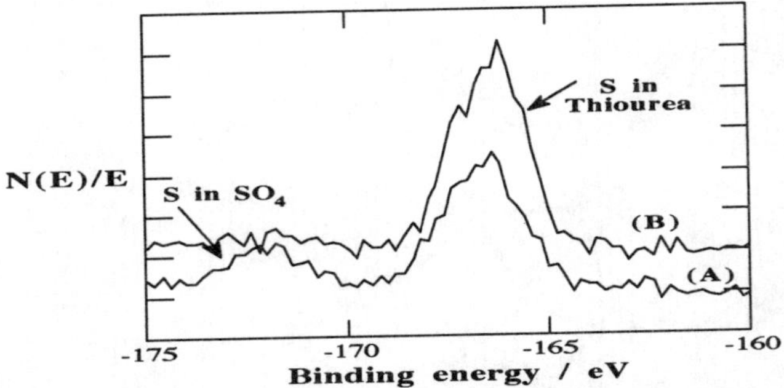

Figure 3. Multiplex XPS spectra of the S 2p peaks in the sulfate (A) and chloride (B).

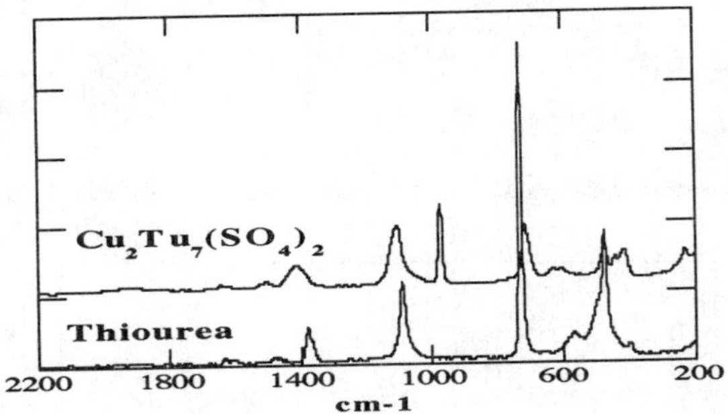

Figure 4. FT Raman Spectra of the cuprous thiourea sulfate and solid thiourea.

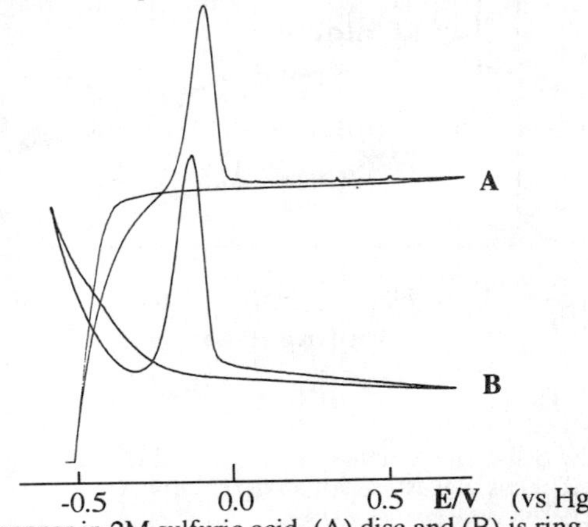

Figure 5. RRDE CV for copper in 2M sulfuric acid, (A) disc and (B) is ring response.

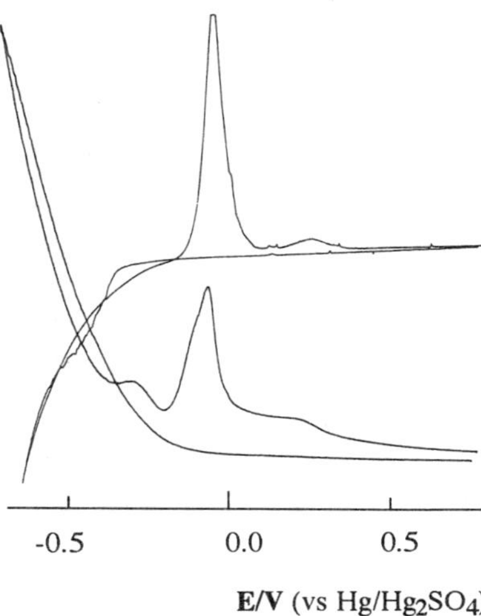

Figure 6. An RRDE CV for copper in 2M sulfuric acid with 0.25 ppm added thiourea

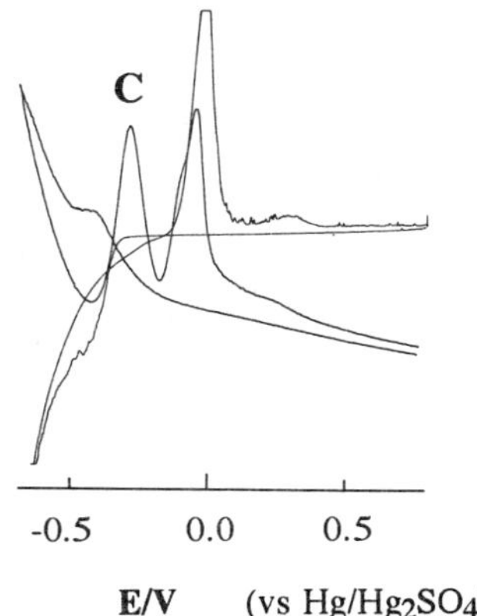

Figure 7. An RRDE CV for copper in 2M sulfuric acid with 0.5 ppm added thiourea. The peak on the ring marked C was obtained during the cathodic to anodic disc sweep.

193

Figure 8 An optical photograph of the macroetched cross section of a commercial copper cathode sample illustrating the effect of plating variations during deposition on the plate structure.

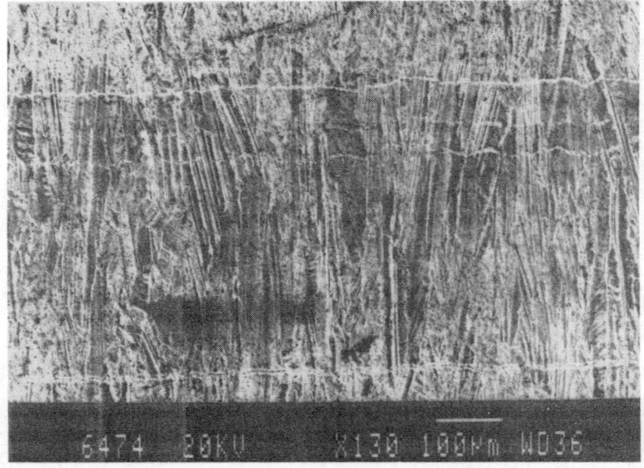

Figure 9. SEM micrograph of a electrorefined copper cathode which has been subject to periodic power interuptions during production. The fine bright lines which traverse the copper crystals were caused by enhanced etching of the plate which was deposited following the re-commencement of plating.

QUALITY ASSURANCE OF MITSUI COPPER FOIL

A.Asano , N.Takahashi , Y.Kohbuchi

K.Yoshimura , H.Kon

Mitsui Mining & Smelting Co.,Ltd.

656-2 Kamakurabashi,Ageo-shi 362

JAPAN

Mitsui Mining & Smelting Co.,Ltd.,as a general
manufacturer of nonferrous metals , has been
accumulating various technologies in both hard
and software fields. We have realized a stable
supply of a number of materials including base
metals , such a zinc, lead, and Copper. As for
copper foil,among other materials,we have been
able to provide products of the world's highest
quality and quantity. Our production flow and
quality assurance system are discussed for the
electrodeposited copper foil.

INTRODUCTION

We have been promoting globalization of the copper
foil business with our main production base and technical
center in Japan, and overseas plants in U.S.A, Taiwan,
France and Malaysia.

THE USE OF COPPER FOIL

Copper foil is one of the most important materials for printed circuit boards which are used for electronics equipment and components.

THE KIND OF COPPER FOIL

Copper foil for printed circuit boards is divided into mainly two types, electrodeposited foil and wrought foil, based on the production method used.

From its combination with non-conductive clad material, the foils are divided into rigid PCB's and flexible PC's.

Copper foil is selected freely. Final product manufacturers and intermediate component producers (laminaters, etchers etc.) select copper foil depending on the application, the processing methods and cost.

THE FLOW OF COPPER FOIL PRODUCTION

Generally speaking, the mechanical properties (elongation, tensile strength and the basic roughness of surface) and the thickness of the electrodeposited copper foils are decided during the electrolysis process.

Other properties such as bond strength, anti-oxidation, printability, etchability, solderability, chemical resistance and humidity resistance, etc. are decided during the surface treatment process. (Production flow sheet is on next page.)

In the case of wrought copper foil, the former mechanical properties and thickness are produced in the rolling process.

QUALITY ASSURANCE SYSTEM

IEC(International Electrotechnical Commission) and IPC(The Institute for Interconnecting and Packaging Electronic Circuits) have official specifications for copper foil and copper clad laminate.

The specification of the copper foil is determined by each customer's requirements while passing IPC specification.

(1) POLICY

Initially, we need to have a thorough knowledge of our customer's specifications and requirements.

Based on those needs, we defined the quality level of our products. A production system which gives stable characteristics is then devised to meet those requirements.

Fig I   THE PRODUCTION PROCESS OF ELECTRODEPOSITED COPPER FOIL
Scrap
Cu
Dissolving
tank
Solution tank
Row foil
Electrolytic Cell
Water Recycle System
Row foil
Treater
Treated
foil
Slitter
Shipping

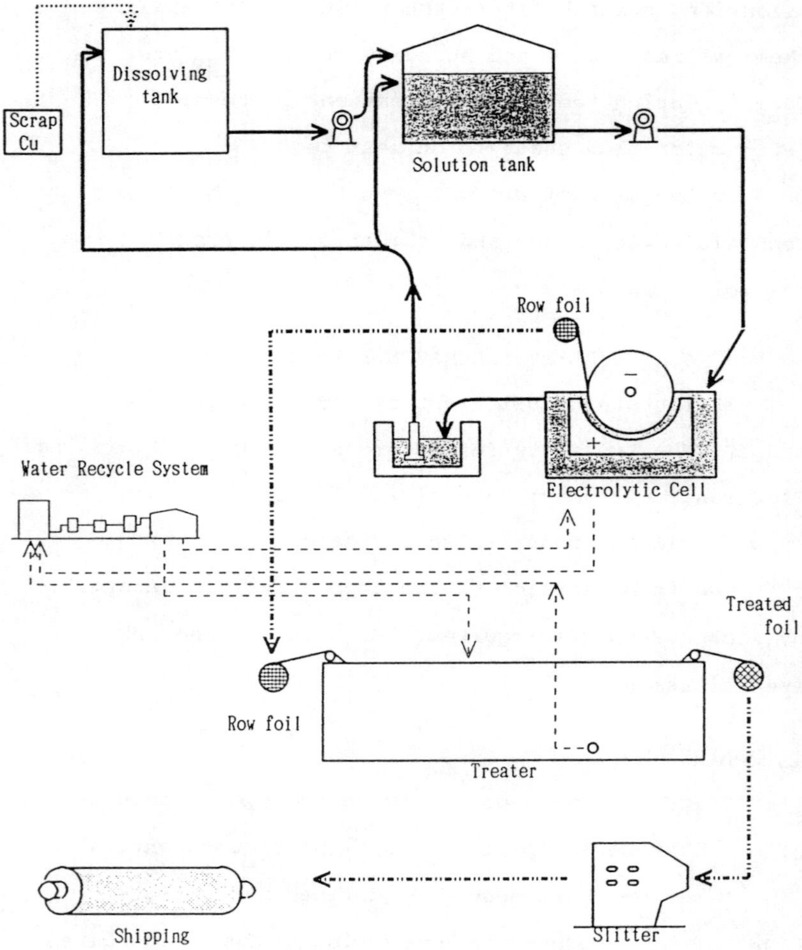

<u>(2)OUTLINE OF QUALITY ASSURANCE SYSTEM</u>

A) Sales department which involves the interaction between individual customers and our company.

B) Quality assurance department which has the following 3 functions.

<u>First function</u>

a) Give detailed understanding of customer's requirements and conformation of specifications to customer's needs (this includes the inspection method).
Establish internal and assuring standards conformation to those specifications up to manufacture of the first product.

b) Assure the quality of the shipped products.

c) Give technical support including the handling of customer complaints.

<u>Second function(Inspection)</u>

Accomplish inspection which is based on the given specification.

<u>Third function(Quality assurance system)</u>

a) Ensure the correct application of quality assurance theory in the quality control.

Fig Ⅱ The information flow in quality assurance system

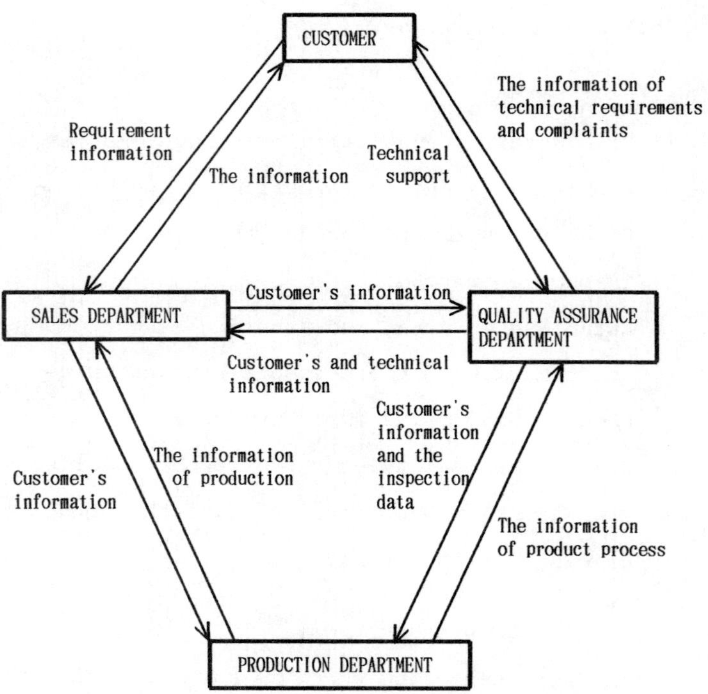

CUSTOMER
The information of technical requirements and complaints
Requirement information
The information
Technical support
Customer's information
SALES DEPARTMENT
QUALITY ASSURANCE DEPARTMENT
Customer's and technical information
Customer's information and the inspection data
The information of production
Customer's information
The information of product process
PRODUCTION DEPARTMENT

Overall Quality assurance system outlined
above.

b) Handle quality audit.

(3) <u>THE 3 ROLES OF THE PRODUCTION DEPARTMENT</u>

The production department has 3 roles in our quality
assurance system.

a) Establishing production conditions which meet
the company's internal standard.

b) Establishing process specifications and manual
(including job training).

c) Appointing those in charge of quality control
and defining their authority and responsibilities
(involving process control and conformation of
company internal standard).

REFERENCES

1. IEC standard, Metal-clad base materials for printed
circuits(Publication 249-2,249-2A,249-3A etc.)
International Electrotechnical Commission, Geneve
Suisse.

2. IPC STANDARD, Metal Foil for Printed Wiring
Application(IPC-MF-150F), Test Method Manual
(IPC-TM-650).
The Institute for Interconnecting and Packaging
Electronic Circuits, Lincolnwood, IL.

QUALITY CONTROL IN COPPER FOIL PRODUCTION
AT NIKKO GOULD FOIL CO., LTD.

Yoshitaka Taniguchi
Takashi Suzuki
Tsuyoshi Konno
Nikko Gould Foil Co., Ltd.
Hitachi Plant
3-3-1 Shirogane-cho
Hitachi, Ibaraki 317 JAPAN

Nikko Gould Foil Company Limited was established as a joint venture between Nippon Mining Co., Ltd., Japan's leading nonferrous metals and petroleum producer, and Gould Inc., the world largest copper foil manufacturer based in the US, in 1981.
The company started its high quality copper foil production at Hitachi Plant in 1982, using electrolytic forming and surface treatment technology. Our two major products, the electro-deposited foil for glass-epoxy printed circuit boards and the treated rolled copper foil for flexible printed circuits, have been enjoying their first rated reputation in the world market, owing to our highly standardized quality control system.
Our production flow, quality management system, and some quality control examples are discussed using the electro-deposited foil, which is mainly used for glass epoxy printed circuit board application, as an example.

INTRODUCTION

In recent years remarkable progress has been made in the electronic industries especially in the field of computers and telecommunication, aiming at high performance, multi function, and miniaturization.
The star player of this technical improvement is the IC tips. However, the leading player behind the scene can be the Printed Circuit Boards (PCB) on which IC and other electronic devices are mounted.
Copper foil, the basic material used for PCB, is continuously required to achieve better quality in its metallurgical properties, surface morphology and characteristics, meeting the customers severe needs.
In this paper, information on our Electro-Deposited Foil (EDF), its properties and production flow are described first and our quality control system to achieve these requirements is discussed.

FOIL APPLICATION AND STANDARDS

EDF for PCB Application

PCB is classified into two types, Rigid PCB and Flexible Printed Circuit (FPC).

EDF is mainly used for Rigid PCB and Treated Rolled Copper Foil (TRCF) is specially used for high ductility FPC.

Copper foil is laminated with non conductive resin substrate at elevated temperature to get Copper Clad Laminate (CCL). A circuit pattern is printed with a etching resist on the copper surface of CCL, then an unneeded part of copper is etched off letting the remaining copper form electrical circuits for PCB.

<u>Foil Standards</u>

Standard specifications of the copper foil for PCB are determined in IPC-MF-150F for both EDF and TRCF.

In this IPC standard, EDF is categorized in four grades, standard foil, high ductility foil, high temperature elongation foil, and annealed foil.

Also in this specification, EDF physical properties are determined by the foil thickness, 18, 35, and 70 micron. In addition to these standard thickness, the very thin, 12 micron or less, foil and others like 25, 50, 105 micron foils are commercially provided.

ELECTRO-DEPOSITED FOIL PRODUCTION FLOW

EDF production flow consists of three steps, electrolytic forming, surface treatment, and slitting. Schematic diagrams of the electro-deposited foil production are shown in Figure 1.

<u>Electrolytic Forming</u>

The raw material, the copper wire scrap, is oxidized to be dissolved into copper sulfate and sulfuric acid electrolyte in the digester. Copper is electrowon continuously on the rotating cathode drum, the chromium plated stainless steel drum, in the electro-forming cell. The deposited thin layer of copper is continuously stripped off and wound into a roll to get the base foil.

The foil has two surfaces; a shiny side which attached to the drum surface is smooth and is a replica of the drum surface, and a matte side which grew into the electrolyte by electrolytic deposition is rough and is characterized by crystal growth. During CCL and PCB production, the resin substrate is laminated to the matte side and the etching resist is printed on the shiny side.

In this electroforming stage, the metallurgical properties, such as tensile strength and elongation, and surface morphology of the foil are determined by the electrolytic conditions.

The foil thickness is controlled automatically by the current applied to the cell and the rotating speed of the cathode drum. The sub weight variation of the foil is computer controlled using our sophisticated thickness control system in the electroforming cell.

<u>Surface Treatment</u>

With the treater, the base foil surfaces are modified in three major stages. All the treatments are done electrolytically at high speed under different conditions. The additional properties required for PCB application are given in these stages.

<u>Nodule treatment.</u> In order to give higher bond strength to the resin substrate after lamination, copper nodules are electrodeposited on the matte side with copper sulfate

and sulfuric acid electrolyte at a high current density close to the limiting current density. The nodules give anchoring effect on bonding.

Thermal barrier treatment. To prevent thermal decomposition of the epoxy resin, which is catalyzed by copper and results in poor bonding after lamination, the brass, Cu and Zn codeposition, layer is formed on the matte side using a cyanide bath.

Stabilization. Zinc chromate stabilizer layer is electro-deposited on both the shiny side and the matte side in order to prevent oxidation of the copper surface by humidity and high laminating temperature.

The matte side SEM pictures of the base foil and the treated foil are shown in Figure 2. A schematic drawing of the cross section of the treated foil is shown in Figure 3, illustrating the layers given by the treater.

Slitting

The treated foil is then slit into the specific sizes requested by the customer ready for shipment.

FOIL PROPERTIES AND REQUIREMENTS

Metallurgical Properties

Metallurgical properties of the EDF are coming from the structure of the copper crystal in the base foil. The major properties are listed below with the numbers for our 35 micron standard foil.

Ultimate Tensile Strength	: 34 - 37 kg/mm^2
Room Temperature Elongation	: 14 - 16 %
High Temperature Tensile	: 20 - 23 kg/mm^2
High Temperature Elongation	: 2 - 3 %
Matte Side Roughness (Ra)	: 1.2 - 1.4 micron

Additional Properties

In the above mentioned IPC standard, only the metallurgical properties of the foil itself are determined. However, taking the PCB production process and its application into account, the foil properties combined with resin system and PCB processing technique are also important (1). These additional properties, such as bond strength to the resin and etching characteristics, can vary with the type of the resin and the processing conditions. Therefore, they can not be determined by the copper foil alone.

The principal supplementary properties to be considered are listed below.

Peel Strength	: Bond strength measured by 90° peeling
Peel Thermal Aging	: Bond degradation after thermal treatment
High Temp. Anti Oxidation	: Anti oxidation at lamination temperature
Printability	: Etching resist adhesion
Etchability	: Facility in etching
Solderability	: Wettability with molten solder

<u>Current PCB Technology Trends and Foil Property Requirements</u>

Owing to the recent trends of the electronic industries, PCB is heading the technology directions listed in Table I. Based on these directions, current requirements of the copper foil properties are also summarized in Table I.

Table I. Current PCB Technology Trends and Foil Property Requirements

Technology Trends	PCB Properties	Foil Property Requirement
High Density Packaging Finer Line Circuit	Good Etchability Accurate Circuit	Low Profile Printability, Etchability, Solderability
Thin Core Multi Layer	Foil Crack Free	Good High Temperature Elongation
High Performance Resin System	No Oxidation through High Temp. Lamination Process	High Temperature Anti Oxidation
High Reliability Zero Defect	No Circuit Exfoliation No Short Circuit	High Peel Strength Cleanliness — No Foreign Particle
Cost Down — Automation	No Foil Wrinkle while Processing	Handlability — High Tensile

QUALITY CONTROL SYSTEM

<u>Quality Goal</u>

Our goal is to fully satisfy our customers through our products and service. Our quality control system is the principal measure to pursue this goal.

<u>QC Organization</u>

Quality Assurance Section is the control center of our QC activities and is independent of the Production Department. This section consists of two sub sections; Quality Assurance, and Inspection & Chemical Analysis. Customer Service is included in the Quality Assurance sub section.

Besides the regular QC organization, the Quality Control Committee, chaired by the plant general manager, examines the overall QC performance and determines the updated plant's QC policy and the term objectives. Following the current QC objectives, a energetic and comprehensive project work to implement ISO-9000 standard is now in progress.

Statistical Quality Control

$\overline{X}$-R Charts and other statistical QC methods are used for process and products control. Once the inspector find a irregular data exceeding the control limit, the corrective response is immediately taken by the production section following the action flow chart.

Production Control

More emphasis is put on the process control to ensure the high quality products. Our steady production is operated following the Standard Operating Procedure, SOP, based on our electrochemical technology and experience. The SOP is periodically reviewed and updated.

Furthermore, automatic process control systems and in-line inspection equipment are widely used in our production. Their performance is also checked periodically tracking the inspection standard.

QC Education

In order to assure that our quality control system is well understood and properly operated, a series of quality education programs are conducted vigorously around the plant. The contents and the level of programs are designed for the class of employee, from the elementary level to the management level.

In addition, QC circle activities are an important measure for the employees to participate in our QC systems and learn practical QC methods. Now 20 circles are actively working on their specific subjects in the plant.

QUALITY CONTROL EXAMPLES

The following two examples would well explain how our quality control works in the electrochemical foil production process.

Metallurgical Property Control at Electroforming

The metallurgical properties of the foil are determined in the electroforming process by the electrolysis conditions such as current density, bath chemistry, bath temperature, additives, and so on. In order to maintain the foil properties within the desired level, every electrochemical parameter is properly controlled.

Current Density. Rectifier output current determines the current density and is set to maximize the production efficiency. Because the current applied to the cell and the cathode rotating speed determines the foil thickness, these parameters are computer controlled and monitored continuously.

Bath Chemistry. Copper and sulfuric acid concentrations are the two major parameters affecting the foil properties. Using the on-line analyzer, copper and acid concentrations are measured and automatically controlled. In addition, a certain amount of the electrolyte is sent to the electrolyte purification plant to avoid condensation of the impurities.

Additives. Organic additives are used to control the nucleation and crystal growth of copper deposition, as a result, to determine the metallurgical properties and surface morphology of the foil. Because the other electrolysis parameters are maintained automatically without big difficulty, it can be said that the precise control of the additives is the key to the consistency of the foil quality.

Usually the additives are controlled by keeping their addition rate to the electrolyte constant, for instance, added in the unit of grams per ton of copper deposited in the cell. Since the additives decompose in the electrolyte by hydrolysis and electrolysis, it is difficult to maintain the "effective" additive concentration at a desired level. Although some analytical methods are reported to measure gelatin concentration in the copper electrolyte, these procedures might not give us the "effective" concentration and sometimes not be practical for taking longer analysis time (2).

In the electroplating industries, Cyclic Voltammetric Stripping (CVS) method is used to evaluate organic additives in the electrolyte (3, 4). Based on the CVS technology, we have developed an electrochemical method to analyze the "effective" additive concentration in the copper foil electrolyte. This method have been proven to be satisfactory through our operational experience. Controlling additive supply rate based on the CVS measurement allows us to produce the high quality foil with consistent metallurgical properties.

Quality Assurance. Mechanical properties of the foil are measured and inspected every production roll. It can be considered this is just to make sure the electroforming process is controlled properly.

Surface Quality Control

Surface Appearance Properties. Recently customer requirement for the appearance of the foil surface is getting severer, besides the regular properties, such as peel strength, high temperature anti oxidation, and printability, etc. The surface appearance properties include small stains, pits and dents, scratches, uneven nodules on the matte side, and small foreign particles ,both conductive and non conductive, on the shiny side. Though these defects are very small, less than 100 microns in diameter, they can be serious defects for finer line PCB application, especially 100 microns or less in line width. The high end PCB producers, who produce PCB for the super computers, first introduced Automated Optical Inspection (AOI) system to detect very small defects about 50 micron in diameter. Now the installation of AOI system is getting popular in PCB industries.

In-line Inspection at Treater. In order to meet the customer requirements and the inspection level, we introduced the in-line optical defect detectors with CCD and computerized image processing technology in 1987. The equipment has similar sensitivity to AOI and checks the surface appearance of all the products passing through the treater. The inspection results are printed out and examined, then sent to the slitter for their information. Moreover, the laser beam pinhole detector is equipped in the line for continuous inspection of the small pinholes on the foil.

Environment Control. To prevent surface contamination after treatment, the slitters are located in the clean room separated from the wet area where the electroforming cell and the treater are found. The slitters are specially made for clean room use and equipped with strong magnetic web cleaners.

Quality Assurance. In addition to the in-line inspection, the final eye inspection is made for every customer roll in this clean room for quality assurance.

CONCLUSION

High quality copper foil for PCB application is electrolytically produced under strict quality control at Nikko Gould Foil Co., Ltd. Statistical process control and in-line inspection enable the sophisticated quality assurance. Moreover, the ongoing extensive work to implement ISO-9000 standard is believed to further refine and reinforce our quality control system to achieve our goal.

REFERENCE

1. H. Daiguji, J. of the Mining and Materials Proc. Inst. of Japan 107, 909 (1991)
2. N. Koura, A. Tsutsumi and K. Watanabe, J. of the Surf. Fin. Soc. of Japan, 40, 317 (1989)
3. R. Heak, C. Ogden and D. Tench, Plating & Surface Finishing, 68, 52 (April 1981)
4. R. Heak, C. Ogden and D. Tench, Plating & Surface Finishing, 69, 62 (March 1982)

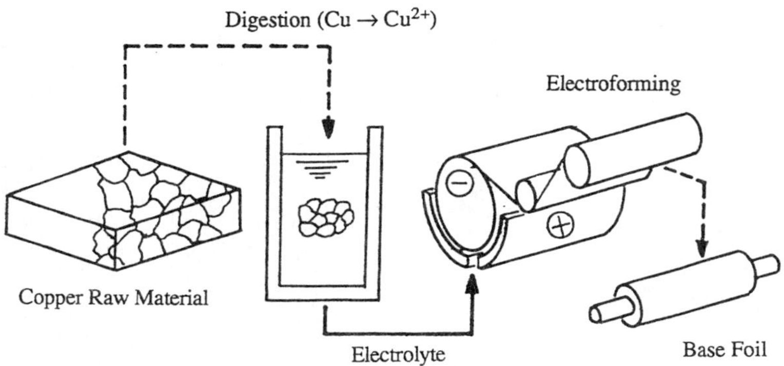

(a) Electrolytic Forming

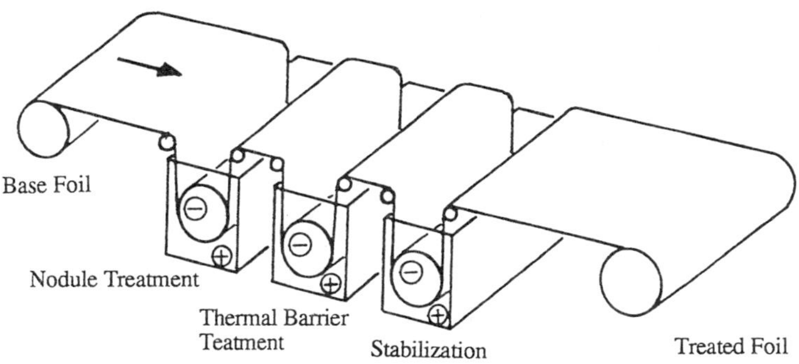

(b) Surface Treatment

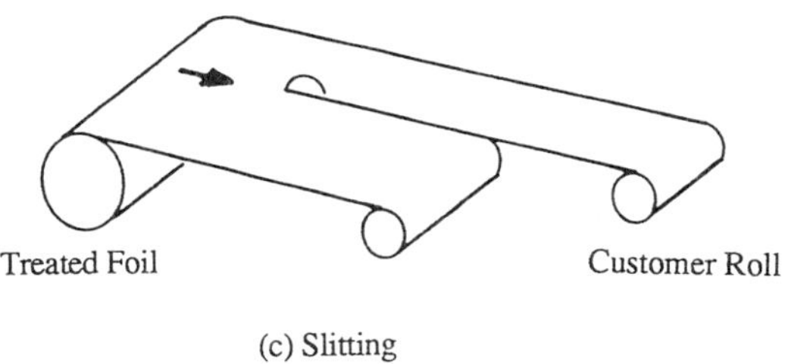

(c) Slitting

Figure 1. Electro-Deposited Foil Production Flow

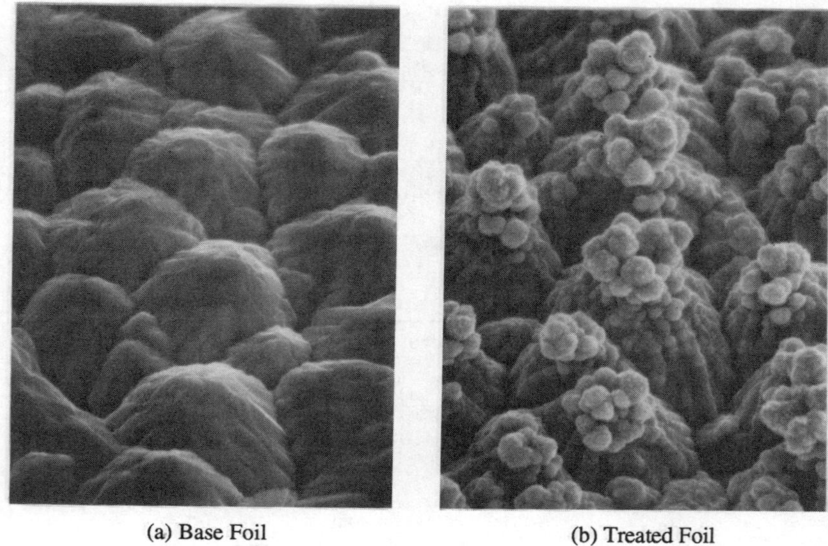

(a) Base Foil   (b) Treated Foil

Figure 2. SEM Image of the Matte Side, 35 micron Standard Foil (×3000)

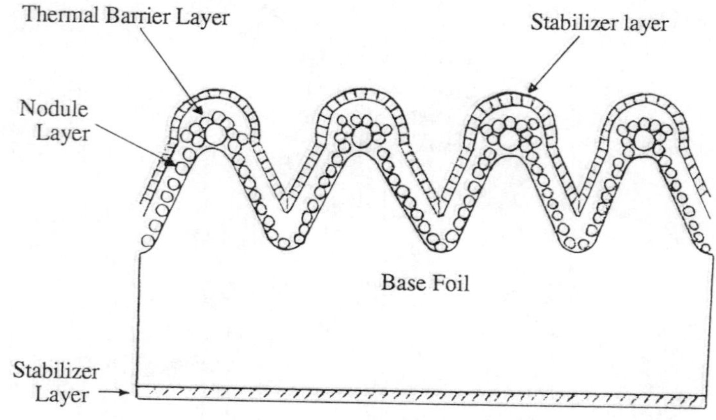

Figure 3. Schematic Cross Section of the Treated Foil

210

CONTINUOUS IMPROVEMENT IN MAGNESIUM

C. B. Wilson,
Magnesium Technical Manager
The Dow Chemical Company
Freeport, TX 77541
ASQC CQE #21519, Member TMS

You may not be able to teach an old dog new tricks, but you can bring new management techniques to an old business. The Dow Chemical Company has been in the magnesium business for 75 years and that is longer than any other company. But that does not mean that the business cannot undergo dramatic change and improvement in the management style of the business. The starting point is a dedicated manager and a dedicated management team. This then extends down through the organization with dedicated and trained workers taking part in the continuous improvement process at all levels of the organization. Key elements of the process are supplier involvement, cross functional interaction, customer focus, and the use of statistical techniques.

INTRODUCTION

The Dow Chemical Company first started producing magnesium metal in Midland, Michigan, in 1918, about 75 years ago. The production processes have undergone significant metamorphosis during this period of time from the crude cells that produced only a few pounds per day to the modern cells that run in excess of 200,000 amps in load and produce over two tons of metal per day. Although there is a lot of interesting process history here, we are going to discuss some of the process management techniques used today in the production of magnesium metal instead of the process itself.

Continuous improvement must be an ongoing process rather than one specific program that runs for a period of time and then stops in favor of the next thing to come in vogue. As such, a fundamental mindset change is required by all the people involved in the organization starting with management who must drive the change. We believe that from a process standpoint we have had a philosophy of continuous improvement for many years, but from a management standpoint "business as usual" seemed to be more or less the way of life.

So you may understand the changes that have taken place in the organization, I will break the system down into four main areas and discuss each of them separately. These four areas are much like the four wheels of a car. If one of them is missing the car just doesn't perform the same. Other wheels can be added for special purposes, but the basic four are still needed. The four wheels of our continuous improvement car are Management Commitment, Employee Involvement, Customer Focus, and Statistical Techniques.

MANAGEMENT COMMITMENT

The first wheel on our car is Management Commitment (See Figure 1). This can be considered one of the wheels needed to steer the car. I first started training on quality improvement systems and Statistical Quality Control (SQC) techniques about 10 years ago. The one constant that all the programs had and that all the "gurus" agreed on is that nothing very positive will happen in the organization unless you obtain management buy-in, commitment, and leadership up front. The first problem which many organizations face is that of giving lip service with little commitment. The second problem is that managers, like most of their employees, don't always catch the vision or understand the message the first time it is discussed. Quality and continuous improvement messages like many others in the business world must be repeated often before everyone understands. In this regard the people in my company are probably not very different from the people in your organization.

What has happened specifically is this. The upper management of our company began to push quality and continuous improvement at the highest levels of the company about 10 years ago. As the programs and the understanding evolved over time, the quality philosophies have slowly but surely permeated the organization. Our management organization in Magnesium Production has been fortunate to have had several leaders who were dedicated from the outset to making continuous improvement a way of life in our business. The manager along with a strong management team has been instrumental in bringing the continuous improvement philosophy to all levels of the organization. Some of the basic tenets of the management team are as follows:

Team-work Needed

The management team needs to work together as a team to make this happen. This means developing mission and vision statements for the management team, and the management team attending something we call a "Quality Performance Workshop" (QPW) as a team. This was followed by a "Quality Performance Workshop/Train the Trainer" session where the management team was trained to facilitate the workshop so they could be trainers for their organization. The QPW combined instruction on the use of basic tools with deployment and team building exercises.

Consistent Management Action

The management team must "walk the talk." This means that the management team must be the leaders in the organization in practicing the tools of quality performance as well as demanding it from their organizations. An indication of how this is working can be gained from the fact that over the past year or so about 75% of the management team's staff meeting time has been spent working on staff-related continuous improvement processes. Projects have been identified that can only be worked on at the management level, and teams have been formed to work on these projects.

<u>Quality Planning a Must</u>

Quality planning has to be a way of life in everything we do. Many activities fall under this area. For instance, all groups in the organization are expected to have mission statements that are developed and owned by the group. A vision of the future is formulated by the management team and communicated throughout the organization. Policies and goals are formulated with a customer focus. Continuous improvement is emphasized in the policy and planning documents. Quality improvement tools are used to identify and prioritize quality improvement issues. The use of improved meeting management tools and facilitator techniques has dramatically improved the quality of meetings and saved an immeasurable amount of time by improving the productivity and focus of the meetings. This has also resulted in a focused problem-solving approach to translate Quality Performance issues into defined projects with individual and/or team accountability. The management team, for instance, spent a significant amount of time developing a long list of things that could be addressed as breakthrough items to reduce costs and make the business more profitable. The list was categorized, and then pared down to a few "doable" projects in each category. Individual sponsors were then assigned to each project and teams formed to carry out the projects. Key performance indicators were then assigned to measure the progress, and a time frame for the improvement. Periodic reports on the progress are made back to the entire Magnesium Leadership team.

EMPLOYEE INVOLVEMENT

The second wheel of our continuous improvement car is Employee Involvement (See Figure 2). Management must be committed, but if the employees are not involved in the process, the effort will be stifled. Employee Involvement covers four key areas: Commitment by All, Cross-functional Interaction, Supplier Involvement, and Process Focus.

A key area that must be addressed to obtain commitment by all is employee training. As managers, we can hardly expect our people to buy into new concepts if we don't provide the training on the concepts. To this end, we launched a significant training effort to ensure all the employees went through the Quality Performance Workshop. This was significant because the managers were the primary facilitators of the workshops and the employees participated with the other people in their Natural Work Teams. Another aspect of this was that the first line supervisors, who were often a part of more than one Natural Work Team, often attended several sessions of the workshop. This had the effect of further strengthening the lessons being taught.

We have had training plans and employee development plans for a long time geared for our supervisory personnel, but with recent changes in philosophy this effort is now being extended to all other personnel as well. This has led to a much higher level of involvement of all personnel in generating mission statements and in goal setting and planning. Here you can see some examples of Mission Statements generated by various groups within the organization (See Figure 3). This higher level of participation is also exemplified by the participation of the hourly folks on numerous cross-functional teams. In fact, the mission statement for Magnesium Operations as a whole was put together by a team composed mainly of hourly employees. Also several hourly employees visited a customer's plant to discuss a quality problem the customer was having, while still others visited a customer's plant to receive a quality award.

Cross-functional teamwork is evident in all levels of the organization. Teams composed of sales, marketing and plant personnel regularly visit customer plants to discuss product, service and quality. These team/customer interactions have allowed us to meet many customer concerns which were previously unknown to us. Interdepartmental teams have allowed us to address problems that are common across departments and implement improvements. Cross-functional safety teams are a mainstay of our safety effort. The Magnesium Department safety efforts are judged by the same criteria as the rest of the chemical complex, and Magnesium results stack up very well in this competitive arena.

Another aspect of this involvement is in dealing with suppliers. There is a strong supplier involvement and evaluation process that is used to enhance the role of the suppliers. Supplier teams deal with suppliers of everything from graphite and alkalinity to labor services. These teams meet with the suppliers and discuss both performance and how the supplier can enhance our production process. Suppliers use statistical methods to monitor and improve the performance of their processes; data and measurements are mutually shared with the suppliers.

One of the fundamental ways to keep employees involved is to maintain a strong focus on the process. Even though people, times and programs change, the process is still there. Critical processes, both production and management, need to be identified and flowcharted. These flowcharts allow everyone to focus on the process rather than being drawn into emotional issues. The use of flowcharts has been a great aid in simplifying operations. These then lead one to the points where measurements are needed in the process. Using statistical techniques to analyze the measurements provides guidance for variability reduction. Where solutions are not clear, group voting and prioritization techniques help to bring the focus back to where it should be.

CUSTOMER FOCUS

The third wheel of our continuous improvement car is Customer Focus and satisfaction (See Figure 4). Without a solid base of satisfied customers, no business can exist for long. Even though we are in what one might consider a commodity business, we still work very hard at keeping our production focus on meeting the customers' needs and, to the maximum extent possible, to differentiating our product in the marketplace. For the customers of primary magnesium, this means producing a variety of product shapes at the weights that the customers desire. Weight control of a large ingot may not seem very important, but many of our customers use our material as an addition to another metal to make an alloy and precise ingot weight becomes a major issue in these situations (See Figure 5). If the customer is adding a 500 lb. ingot to a melt, they want it to be 500 lbs. and not 400 lbs. or 600 lbs. Through some of our cross-functional meetings with the customers, we discovered just how critical weight control was and also that we needed to do some work to improve our situation.

After some diligent work by the supervisors and operators in the casting operation, we were able to make a significant improvement in our performance. These improvement efforts continue with statistical monitoring of the process to make sure where we are operating on an hour-to-hour basis, and periodic statistical tabulation of the data to examine how the large data bases look. A further indication of this dedication to the quality of the product is the Continuous Vertical Direct Chilled Caster which is being installed to provide improved product quality to our customers. This plant is scheduled for start-up in the next few months. We also track communications which come back from our customers. This data is categorized into several categories. The feedback is encouraged so we have the opportunity to fix anything that is wrong with the system and zero in on any problem areas which we may have.

STATISTICAL TECHNIQUES

The fourth wheel of our continuous improvement car is the use of Statistical Techniques (See Figure 6). I firmly believe that if anyone is going to make much sense of the data coming out of any process or system, that statistical data analysis techniques must be used to analyze the data. Many people interpret the use of statistical techniques as the use of SQC charts and maybe some design of experiments. We have taken the tactic that there are a lot of statistical techniques out there, and we need to apply statistical thinking to the process and the process data at appropriate points and use only those that make sense. The other thing we have found is that because of the electronic controllers which control a lot of our processes, most of the on-line process data are correlated and traditional "X Bar and R" or "X mR" control charts do not always provide useful information. As a result, there are not as many SQC type control charts used in the chemical-end of our business as might be done in piece good or machining type applications. Here are just a few of the applications which we do use:

In the casting area of the process, X mR type control charts are used to keep track of the weights of the ingots coming off of the casting line. This is then coupled with the analysis of some large data base histograms of the ingot package weight data which comes from the warehouse scale data. Within this same system, we also periodically take continuous ingot weight data of the ingots coming off the casting machines to determine the ingot-to-ingot variability. These data points are compared to the results of the other analyses.

In our Denver extrusion plant, we manufacture a lot of things which might be considered piece goods. A variety of statistical techniques are used here to ensure that product quality is met. In the examination of some incoming machined parts, we have switched from the use of go/no-go gauges to other gauges that can provide variable type data. This has allowed us to use a Mil Standard 414 sampling plan which looks at the average and standard deviation of the part. As a result, we worked with our suppliers to improve their quality. This in turn improved the quality of the product we send to our customer.

People often misunderstand statistical techniques. As a result, one can get strange requests. We had an example of this at the Denver operation. One product we were making was not performing too well and required a tedious 100% inspection. We were challenged to come up with a statistical sampling method which would not require 100% inspection. A control chart plotted on the reject data showed the reject rates to be out of statistical control. This analysis led to the conclusion that we had a process problem and not a statistical sampling problem. After identifying and solving the root process problem, it was easy to institute a sampling scheme which would pick up any deviation. This also resulted in a significant productivity increase for this phase of the operation (near 100%) not to mention the savings in scrap rate which were running above 10%.

While control charts are not plotted on all variables, they are used extensively to examine virtually all process data and to aid in determining the variability of the process at various points. We have developed computer plotting interface programs so that all routinely generated numbers from our control labs and many routinely generated process numbers can be plotted in control chart format. Since our process control labs produce close to 1000 individual points per day, this is an awesome array of data to evaluate. With this much data from the labs, not to mention the continuous process data, one has to be selective as to which points are plotted on control charts. We also have other programs which look at the average and standard deviation of all the product analytical data over specified time frames. This is a great aid in calculating process capabilities. Histograms of the product analytical data are routinely produced from the databases. These very large data base histograms are useful in analyzing the process and are also used to present the data to our customers (See Figure 7).

In the electrolytic cells, because of the multiples of cells and the mass of data, it is necessary to be selective about the statistical techniques that are routinely used. Many control charts are plotted around small segments of the process to study what is going on. In one such effort numerous XmR control charts and X Bar and S charts were plotted to study the results of the program used to correct the composition of the electrolytic cells. After combining this with lab variability studies, an Exponentially Weighted Moving Average control chart (EWMA) was devised. A computer program takes the laboratory data and calculates the exponential averages. If either the upper control limit or the lower control limit of the EWMA chart is reached, a response is triggered. This calls either for an addition to the cell or a subtraction from the cell depending on which control limit is hit by the EWMA. The chemistry of each electrolytic cell in the process is now corrected by the control limits of the EWMA control chart on that cell.

Control charts are not limited to normal operations data. Two areas which receive an ever increasing emphasis both because of internal concerns and policies and external regulatory pressure are safety and ecology. Safety statistics are plotted and tracked using a c-chart. The use of statistical charts for this type of data has proven to be a valuable tool because it allows the safety managers and focal points to judge whether an increase or a decrease in the data is something for serious concern or just a part of the normal system. In times past, a single point increase was assigned greater significance than was warranted, and as a result, there was a lot of "playing around with the system." This response to noise in the system often wasted valuable resources which should have been spent in working on root causes. With the increasing amount of government regulations in the environmental area, there are monitoring and sampling requirements which didn't exist just a few years ago. We have been very proactive in this area of our operation. We do extensive control charting and statistical analysis of our environmental monitoring data. It is extremely important for us to understand both averages, standard deviations, and capabilities of both the production process and the emissions data as well as whether or not the systems are in a state of statistical control. Unless good data are generated and a good statistical analysis is done on the data, one would not be in a good position to negotiate permits with regulators.

RECOGNITION

Our magnesium organization is recognized both by our customers and by our upper management as one of the leaders in Continuous Improvement. From our customers we have received a total of 17 quality awards over the past 5 years. These are extremely important to us since customer satisfaction is what it is all about. Internally, the magnesium business was recently recognized by receiving the "Dow Chemical President's Award for Quality." Magnesium was the first production organization in the company to receive this prestigious internal quality award, which is a significant tribute to the many men and women who make up the Magnesium Department.

CONCLUSION

This is a brief look at our Magnesium "Continuous Improvement Car" (See Figure 8). With all four wheels, Management Commitment, Employee Involvement, Customer Focus, and Statistical Techniques on the ground, it is a smooth running vehicle. Get a flat on any wheel, and it slows down not to mention how hard it is to drive. If other tools come along which can help the machine, it is adaptable to receiving them, so long as the basics remain. Indeed I have found that although the magnesium process might be considered an "old dog," it really is possible to teach it some new and exciting tricks.

FIGURE 1 - MANAGEMENT COMMITMENT

FIGURE 2 - EMPLOYEE INVOLVEMENT

MAGNESIUM WAREHOUSING

Mission Statement

We exist to provide a smooth flow of high quality products and services to Internal and external customers through teamwork, commitment, and the continuous improvement process

MAGNESIUM OPERATIONS OFFICE PROFESSIONALS

MISSION

To continuously improve service to all Magnesium Operations employees, the Office Professionals are committed to provide the highest quality administrative support by utilizing teamwork, the quality performance process, the latest technology, and effective communications.

FIGURE 3 - MISSION STATEMENTS

FIGURE 4 - CUSTOMER FOCUS

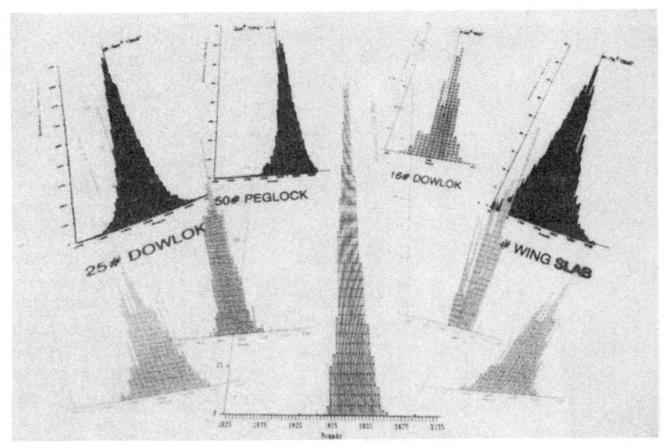

FIGURE 5 - INGOT PACKAGE WEIGHT HISTOGRAMS

FIGURE 6 - STATISTICAL TECHNIQUES

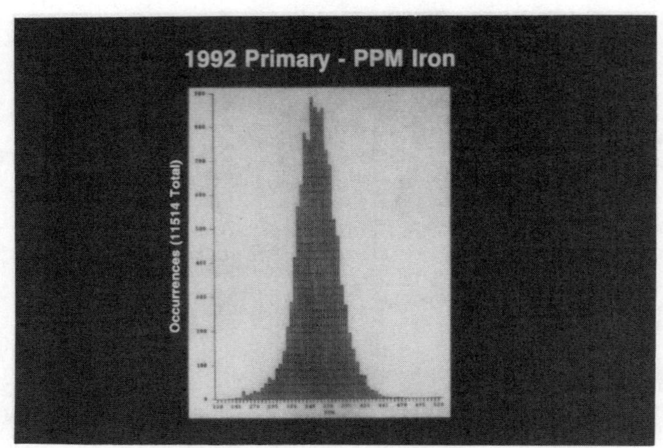

FIGURE 7 - IRON ANALYSIS HISTOGRAM

FIGURE 8 - CONTINUOUS IMPROVEMENT CAR

IMPROVING QUALITY OF ELECTROPLATED PRINTED
CIRCUIT BOARDS THROUGH PROCESS CONTROL

Dennis R. Turner
Consultant
59 Susan Drive
Chatham, NJ 07928

As printed circuit boards (PCBs) become
more complex - higher circuit densities
and multilayered, defects will increase
unless an effort is made to improve processing quality. There are times when the
only solution is to drastically change the
process. It is important to understand
the potential source of defects and what
causes them. Continuous monitoring and
control of critical steps in processing
and operator training are essential to
high yield production of PCBs.

INTRODUCTION

High quality printed circuit boards (PCBs) are important to the success of modern electronic systems. They are
a cost effective and reliable means of interconnecting
electrical and electronic components into a compact system.
The concept of PCBs was invented about the same time the
transistor effect was discovered (1947). PCB technology
developed to meet the needs of electronic systems(1). They
became a perfect match with solid state devices to produce
reliable, compact, low cost electronic systems. Figure 1
shows the segments of the 1991 PCB market applications.

Electroplating is an important part of the manufacture
of PCBs. As PCBs became more complex, with higher circuit
densities and multilayered boards, electroplating defects
became more of a problem. Better quality control of electroplating processes involved in PCB manufacturing became
necessary to obtain acceptable yields at a reasonable cost.
A considerable amount of research and development has gone
into improving electroplating processing capability.
Causes of failure due to electroplating can involve many
factors. It is important to recognize the various types of
defects, understand what causes them, and change the processing to eliminate future defects. Expanding the electroplating process window can be a great help to minimize
defect occurrence.

PRINTED CIRCUIT BOARD TYPES AND CLASSES

Printed circuit boards are made in different types depending on the application and cost.
1. Single sided - SS
2. Double sided - DS
3. Multilayered - ML
SS and DS boards can be rigid or flexible. Thin dielectric SS and DS boards are use to build ML PCBs. Multilayered boards are the most complex, costly, and difficult to make with high quality at high yields. A pie chart of shipments by product category for 1992 is shown in Figure 2. MLBs currently dominate the PCB business and the trend is increasing. The layer count can vary from 3-4 to 24-30 layers. Obviously MLBs with many layers are more difficult to make with good yields and require much greater process control and product inspection.

PCB products can be divided into three classes:
1. Consumer - non-critical, low cost
2. Industrial - high performance, long life
3. High reliability - uninterrupted service
PCBs for consumer products can have high density circuit boards but long life is not essential since the products often become obsolete in less than a decade. Low cost is a major factor. Industrial PCBs must perform reliably under adverse factory conditions - vibration, heat, pollution, and electromagnetic radiation. Ideally boards should be free of defects for the life of the machine or instrument. High reliability PCBs are required in communication systems, computer control and business systems, and military systems where interrupted service is critically important.

THE MANUFACTURING PROCESS

The manufacturing process begins in Planning with the issue of a Job Traveler - see Figure 3(3,4). The Traveler lists all materials and operations as manufacturing instructions. It not only serves as a document to record information on each set of boards produced but it also documents customer requested specification changes.

The manufacture of printed circuit boards involves a large number of processing steps. The complexity of making some PCBs has been compared to the difficulty in making high density silicon integrated circuits. More than half PCB processing steps do not involve electroplating. Each segment of the manufacturing operation must be monitored to insure that acceptable quality processing is passed on to

the next segment. If some preplating steps are not done
correctly, plating problems can occur. The dielectric
substrate, bonded copper, hole drilling, and resist application are potential sources of problems for quality plating. For example, drilling of through holes in MLBs his-
torically has lead to problems for the plater. Production
rates for drilling holes are so fast that the drill bit may
rotate only about 1 1/2 times through the board. Drill
bits reach a temperature of about 500 $^\circ$F (260 $^\circ$C) during
drilling which melts epoxy dielectric and smears epoxy over
the inner copper layer edges of multilayer boards. Epoxy
smear must be removed before attempting to metallize
through holes. This is done chemically using a solution
such as alkaline permanganate which will dissolve or dis-
place epoxy. A dull drill bit will punch more than drill
creating a ragged hole that generally cannot be conditioned
and plated to meet the specification for through holes.

Electrochemical processing steps used in PCB manufacture include: electroless copper plating and electroplated
copper, tin-lead, nickel, and gold. Much of the gold on
contact fingers can be replaced by palladium or Pd-Ni both
with a thin gold flash.

When the design requirements of a PCB exceed the
processing capability of manufacture, then radical processing changes may be necessary to obtain a quality product.
Narrower line widths and spacings require smaller diameter
through holes. This results in higher aspect ratio through
holes that may be difficult to plate with the proper copper
thickness and uniformity using standard electroplating
processes and facilities. Upgrading or replacing equipment
may be required. An electrochemical process change may
also be necessary. Care must be taken to insure that a
process step change is compatible with the rest of the
process. For example, introducing or changing a cleaning
step prior to plating that involves a chemical that would
contaminate the plating solution is a risk if intermediate
rinsing is not effective.

A fully additive (full build) electroless copper
plating process is capable of fabricating PCBs with high
density circuitry depositing copper uniformly across boards
as well as into high aspect ratio through holes(4,5). Electroless plating processes are quasi-stable electrochemical
systems that require more precise control of the solution
chemistry than electroplating processes. Long term bath
stability and physical properties of the deposit are major

problems that must be controlled. Extensive research and
development of electroless copper deposition processes over
several decades has succeeded in finding solution recipes
and additives that can be used reliably for the production
of high density PCBs. Frequent monitoring and control of
the solution chemistry is critical to the success of additive processing.

<u>Impact of Electronic Device and System Designs</u>
     The continual upgrading of electronic device and
system performance in smaller packages has lead to major
design changes that affect PCB design and manufacture. In
general, the trend is toward higher density PCB circuitry.
Also new PCB design concepts such as Surface Mount Technology (SMT) and Tape Automated Bonding (TAB) have been adopted for many applications.

     When higher circuit density PCBs are needed, tighter
quality control is required on electrochemical deposition
processes to achieve an acceptable manufacturing yield. It
may involve the development of improved or new inspection
tests that have the precision and reliability necessary.

     Higher density electronic packages means that compo-
nents must become smaller and interconnections moved closer
together. This requires finer line widths and spacings and
smaller diameter through holes. Adding more layers in a
multilayered board also will increase circuit density but a
thicker board means higher aspect ratio through holes and a
greater challenge to the copper deposition process in the
holes. These PCB design changes tend to favor the use of
additive copper deposition technology(5).

     Another technique used to achieve high density elec-
tronic packages is the "daughter board" concept. The main
PCB ("mother board") is designed to hold various other
small PCBs "piggie back" style. This design is especially
good when the "daughter board" must utilize a different
dielectric such as ceramic for heat dissipation. Creating
printed circuit patterns on ceramic involves a different
technology for developing the initial adherent conducting
layer.

     Surface Mount Technology (SMT) is a means of achieving
higher packaging densities. It eliminates through holes
required to mount components on boards so that finer line
circuits are possible. Also components can be mounted on
both sides of a board. Small diameter metallized via holes
are used to interconnect both sides and interlayers of
circuitry.

Tape Automated Bonding (TAB) involves etching a printed circuit pattern in copper bonded to a thin flexible dielectric such as mylar, attaching a component, and transferring the assembly to the main board. This is a low cost technology that is completely automated.

FACTORS THAT INFLUENCE PCB QUALITY

Various factors influence PCB quality such as the specification, processing capability, delivery time, environmental, and cost. The specification defines board complexity and parameters which must be within specified limits. It determines the potential difficulties toward manufacture at high yields. Generally the specification is written with an eye to processing capability. However, circuit designers can make life difficult for manufacturing engineers by specifying end point requirements that can not be met with existing plant equipment and chemical processing.

In order to be competitive, it is essential that PCB manufacturers continually improve their processing capability. This includes modernizing plant equipment and chemical processing. Most PCB manufacturers rely on outside vendors for the design of equipment and development of new and improved chemical processes. Often the vendor that wins out in competing for equipment or chemical process used by a PCB manufacturer is one that provides the best technical service. Another important factor toward PCB manufacturing processing capability is engineer and operator knowledge and skill. There must be continuing education and training about the manufacturing process and quality standards and inspection tests the product must pass.

Delivery time can influence product quality if the manufacturer has difficulty meeting delivery schedules. Most PCB manufacturing shops know their turn around time for making boards. When delivery is promised in a shorter time to get an order, production quality may suffer when "short cuts" are taken to meet the dead line. Speeding up the electroplating process can be done by raising the plating current density. This reduces the deposit uniformity in through holes and across the board. A manufacturer risks quality by increasing PCB production rate beyond the normal processing capability.

Many of the chemicals used to make PCBs are considered toxic and several are being outlawed for continued use. The consequence is that substitute chemicals must be found

which are as effective. Sometimes this is difficult to do
without reducing processing capability. PCB quality can
decline until the manufacturers learn to cope with changes.
Environmental laws concerning what chemicals can not be
used in PCB manufacture become more restrictive every year.
Proper handling and disposal of waste chemicals is another
problem but PCB manufacturers have dealt with that successfully for several years and many technical conferences on
that topic are held every year.

In a highly competitive market, cost can influence
quality if the PCB manufacturer must economize too much to
stay in business by reducing the work force, avoid modernizing equipment and processing, and/or minimizing quality
control and inspection. Much depends on the management
skill of the PCB manufacturer to know the cost of each step
in processing. Investment in new or modified equipment or
chemical processes must be considered periodically and
weighed against maintaining quality at a competitive cost.

<u>Automation</u>
Automation of PCB manufacture improves quality and
lowers cost only if it is done right and limited to what
makes good sense. The most successful automation systems
are designed and built by people who fully understand what
the machine must do to be successful. Automation can significantly improve a company's competitive position.

QUALITY CONTROL STRATEGIES

The only printed circuit board manufacturers who will
survive are those that develop a broad-based strategic plan
that will produce products that satisfy customers(6). This
requires a commitment to quality at all management levels,
engineers, and production operators. Quality assurance has
the responsibility for satisfying customer quality requirements. Their job is to monitor the entire manufacturing
operation, identify production problems, and see that
problems are corrected. They prepare the written procedures and workmanship standards(2,3).

Quality conformance inspections and tests are done at
various stages of PCB production. As the complexity of the
board increases, more inspections and tests are required.
Figure 4 is test circuitry designed for military PCBs and
are located on the board(s). Coupons are cut out at
intervals of production to test for quality of such things
as through hole drilling, interconnections in multilayer
boards, and plated copper and solder on surface conductors
and in holes.

Statistical quality control is achieved by monitoring
key variables as outlined by the Job Traveler. Plots of X,
R, and P charts determine how well electrochemical processing is performing(7). Deviations from the established
high-low limits signals Quality Assurance and the plating
engineer that adjustments are necessary.

Electroplating Process Control

The quality of PCBs is dependent on maintaining good
process control. The Hull Cell and newly developed hydro-
dynamically controlled "Hull Cell" are useful in evaluating
electroplating processes over a wide range of current
densities and trouble shooting many plating problems(8).
Modern analytical instruments can be more specific about
the concentration of important solution constituents and
the nature of contaminants.

The development of chemical sensors for on-line use
and computer controlled electronics has created an array of
instruments capable of rapid chemical analysis and automatic process control of plating systems(9).

Automatic electroless copper solution analyzers and
controllers are important to the bath stability and metal
deposition rate. Sensing elements used include pH electrodes, specific ion electrodes, polarography, titration,
and colorimetry.

Test Patterns

Test patterns are used to evaluate the capability of
processing to meet quality standards. Figure 5 illustrates
the type of test pattern that can be used to evaluate the
capability of a PCB processing system. There is a tendency
to design circuit boards which require electroplating
processes to operate near the technology limit. Test pattern can establish that limit.

Most Frequent Plating Quality Problems

Electroplating quality of PCBs depends on excluding
defects that might interfere with their proper electrical
function. It is important to understand the potential
source of defects, what causes them, and how they can be
avoided.

Table I ELECTROPLATING DEFECTS - CAUSE AND SOLUTION

DEFECT	CAUSE	SOLUTION
Through holes		
. Corner cracks	Low strength copper	Plating additive improvement
. Voids	Incomplete electro-less copper	Improve hole cleaning & catalyst treatment
. Contact to interlayers	Epoxy smear	Improve epoxy removal
. Copper distribution	C.D. too high, solution flow through low	Reduce C.D. and/or increase solution through holes
. Copper blisters	Excessive board moisture	Check dielectric, post plate bake
. Copper whiskers	Solution contaminant - surface active agent	Remove contaminant, Air or O_2 agitation
Contact "burnoff"	Contact copper too thin, C.D. too high	Increase preplate copper tks., reduce current density
Poor copper adhesion	Copper bonding too low	Increase surface microroughness
Shorts, opens	Masking defects, mechanical damage, dirt, slivers	Check resist application, avoid dirt, metallic slivers
Tin-lead composition	Plating bath not in control	Maintain bath analysis and control
Contact tab defects	Ni, Ni-Pd, Pd, Au baths not in control	Maintain bath analysis and control

Many of the quality problems that occur in the manufacture of PCBs involve through holes(10). Corner cracking occurs when a board is heated during plated solder reflow or in wave soldering of components mounted on boards. Thermal expansion of board dielectric material is greater

than that of copper which stresses the copper at the hole
corners. Strength of plated copper in through holes can be
improved significantly when the right organic chemical
additives are used in copper plating baths.

Voids of copper in through holes can result in open
circuit through the hole and/or no contact to interlayer
copper circuitry. The problem is usually caused by ineffective hole cleaning or Pd-Sn catalyst treatment prior to
electroless copper deposition. Chemical processing before
electroless copper deposition needs to be improved. Unless
there is a continuous electroless copper deposit through
holes, electroplated copper will not bridge voids.

Bad electrical contact between through hole copper
deposits and interlayer circuitry is generally caused by
inadequate epoxy smear removal. The oxidizing agent such
as permanganate solution may need to be reconstituted for
effective epoxy removal.

Copper plating in through holes must be adequately
thick and reasonably uniform(11). If too thin, electrical
resistance through the hole will be too high. Too thick a
copper deposit in prevents component lead insertion.

Copper blistering in through holes is usually caused
by excessive moisture in the dielectric prior to plating.
It is most likely to occur during hot humid weather. A
moderate post plating heat treatment solves the problem.

Copper whisker growth around the openings of through
holes is caused by surface active organic contaminants in
the plating bath(12). The contaminant source is usually
from a preplating alkaline cleaning solution not properly
rinsed from the boards. Whiskers only grow in an oxygen-
free plating solution. Agitating plating solution with
nitrogen gas, for example, should be avoided.

Thin copper contact "burnoff" can occur when the
copper thickness under the electrical contact spots is too
thin to carry the applied current. This is most likely to
happen during high current density copper flash plating
after electroless copper metallization of through holes and
before the lower current density heavy electroplate is
applied.

When a PCB is generated by electroless copper plating
on a bare board, a peel test is usually required whenever
there is any doubt the plating will adhere to a substrate.
Unless there is some surface microroughness for mechanical

adhesion, weak van der Waals adsorption forces are the only means of holding deposited copper to the surface. Swelling and selective etching the bare dielectric with solvents is the usual method of producing microroughness. Copper to copper adhesion failure can occur if the initial copper has a heavy oxide or greasy layer which is not cleaned off prior to copper deposition.

Copper circuit shorts and opens can be caused by faulty artwork, mechanical damage, metallic slivers or dirt particles. A clean quality manufacturing operation will help to minimize this type of defect.

The 60-40 tin-lead solder plate is used as a copper etch mask and for soldering components to a board. It is the desired alloy composition since it melts at the lowest temperature. Maintaining that eutectic composition over long periods of
plating time requires frequent chemical analysis and plating solution maintenance. Copper contamination in the bath will deposit out and cause solder reflow problems.

Contact fingers (tabs) on PCBs are designed to receive connectors. To insure reliable electrical contacts, fingers are gold plated. A nickel underlayer improves contact performance. Gold thickness and hardness are important parameters to be monitored to maintain a long term low contact resistance.

ASSETS TO IMPROVEMENT IN QUALITY CONTROL

Good equipment, well trained operators, and knowledgeable supervisors and managers are essential to consistent manufacture of high quality printed circuit boards. The greatest asset an electroplating facility can have to maintain quality is a group of well trained experienced operators. Training and experience will minimize defects because operators will recognize conditions which need correcting before they become a problem. Well placed sensors can also help to alert operators to conditions approaching out-of-control situations. A sensor must be highly reliable. Sensors that give false signals can be worse than having no sensor especially if the sensor signal is used to automatically maintain a process chemistry or electrochemical function. In critical areas, using multiple sensors or self calibrating sensors is a good method of avoiding improper action from a faulty sensor.

SUMMARY AND CONCLUSIONS

Printed circuit board manufacture can be a complicated process particularly when high circuit density multilayer boards are produced. Considerable planning, high technology equipment and processing, and quality control is required to achieve acceptable yields. The need to make more complex boards has forced electroplating engineers to look for new ways to fabricate circuitry with finer line widths and spacing and higher aspect ratio through holes. Electroless copper plating once a slow and difficult process to control has been improved considerably in recent years and, with modern automatic plating solution chemistry analyzers and controllers, promises to become the standard process to generate PCBs. The industry has developed a sophisticated quality assurance system that makes possible production of quality PCBs in high yields at reasonable cost.

ACKNOWLEDGEMENT

I wish to thank Don Dinella of AT&T Technologies for his advice and suggestions to improve the manuscript.

REFERENCES

1. D.Dinella, "Trends of Printed Circuit Technology in the U.S.A.", 3rd. International Symp. on Printed Circuits, Tokyo, Japan, Nov. 20, 1992.

2. R.H.Clark, Handbook of Printed Circuit Manufacture, Van Nostrand Reinhold, New York, NY, (1985).

3. C.Coombs,Jr., The Printed Circuit Handbook, 3rd ed., McGraw-Hill, New York, NY (1988).

4. H.Nakahara, "Full Build Electroless Copper Plating is the Process of the Future", Electronic Pack. & Prod., p.50-53, Jan. (1991).

5. D.Dinella, "Additive Process Technology for the Future PWB Market", Proc. IPC World Meeting, San Francisco, May 13, 1993.

6. IPC-PC-90, "General Requirements for Implementation of Statistical Process Control", Oct. (1990).

7. AT&T Statistical Quality Control Handbook, p.12 Mack Printing Co., Easton, PA (1984).

8. I.Kadija. et.al., Plat.and Surf.Fin., 78(7), 60 (1991).

9. D.R.Turner and Y.Okinaka, "Automatic Plating Bath Analyzers/Controllers", Amer. Electroplat. Soc., Proc. 68th Annual Tech. Conf., Boston, MA (1981).

10. IPC Tech. Report No. 579, "Round Robin Reliability Evaluation on Small Diameter Plated Through Holes in PCBs", Sept. (1988).

11. D.R.Turner, "A Technique for Evaluating Electroplating into Small Through-holes of Printed Wiring Boards", Plat. and Surf. Fin., 66(7), 32 (1979).

12. P.A.van der Meulen and H.V.Lindstrum, "A Study of Whisker Formation in the Electrodeposition of Copper", J. Electrochem. Soc., 103, 390 (1956).

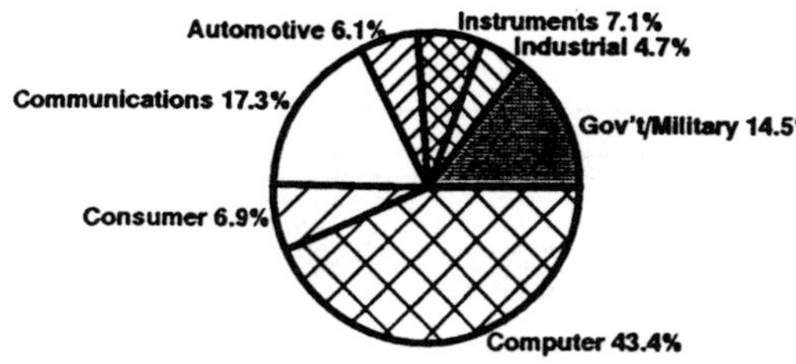

Figure 1
1991 PCB markets

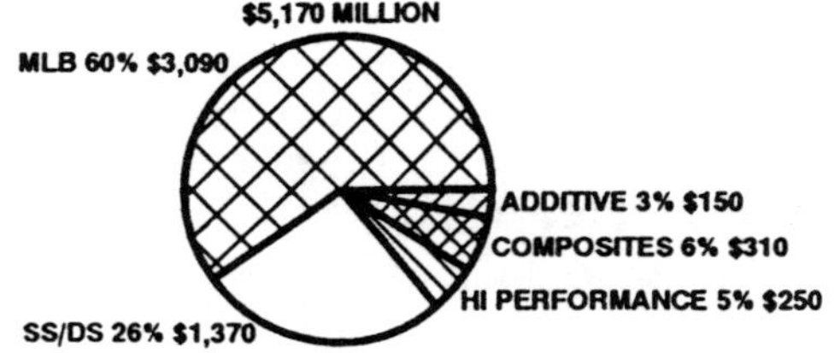

Figure 2
1992 PCB product
category

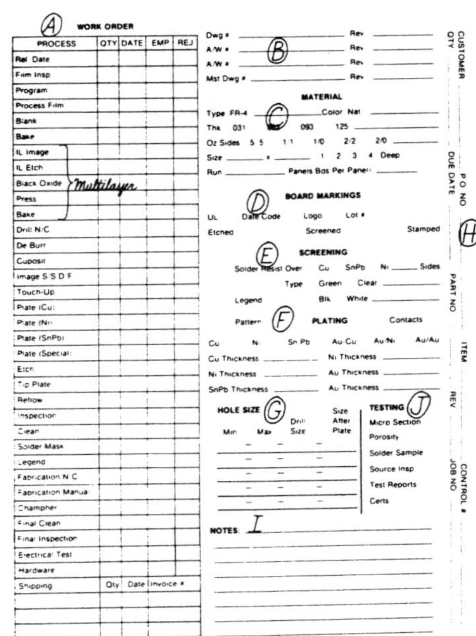

Figure 3   Typical job traveler

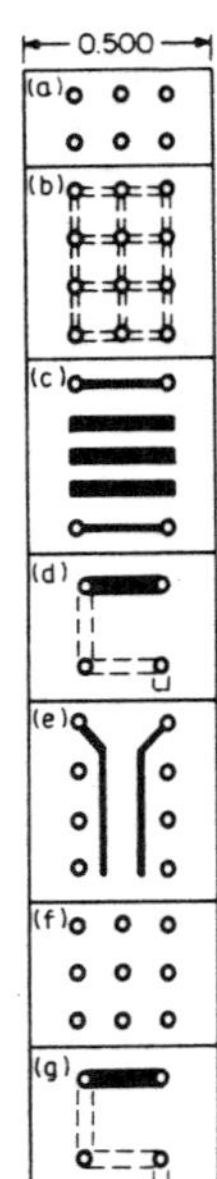

Figure 4   Quality
conformance test
circuitry

235

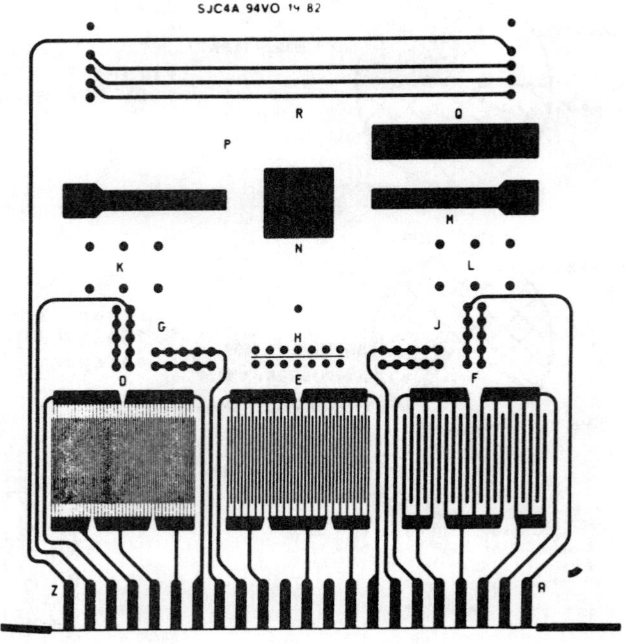

Figure 5   Processing test pattern - IPC-B-25 artwork.

AUTHOR INDEX